Fundamentals of Producing Lift Forces

Tetsuzo Sakamoto

Fundamentals of Producing Lift Forces

From Magnetic Field to Fluids

Springer

Tetsuzo Sakamoto
Prof. Emeritus of Kyushu Institute of Technology, Japan
Kitakyushu City, Japan

ISBN 978-981-92-2221-6 ISBN 978-981-92-2222-3 (eBook)
https://doi.org/10.1007/978-981-92-2222-3

This Springer imprint is published by the registered company Springer Nature Singapore Pte Ltd.
The registered company address is: 152 Beach Road, #21-01/04 Gateway East, Singapore 189721, Singapore

Foreword

The phenomena of lift and levitation have long captured both scientific and popular imagination. In everyday language, "levitation" may suggest magic or illusion, but in engineering it refers to a range of rigorously understood physical effects that allow bodies to counteract gravity. This text embraces this broader perspective, treating levitation as part of a broader family of lift forces and examining how such forces arise in many different physical systems.

Within this unified framework, the author surveys lift and levitation mechanisms. Some are not effective when a body is at rest, as in the case of aerodynamic lift and electrodynamic suspension. But many are effective, as for buoyancy, and acoustic, electrostatic, and electromagnetic suspensions. Each of these modes is introduced and developed with mathematical analysis, giving the reader qualitative explanations and the tools for design and analysis. Drawing upon the author's extensive experience, magnetic suspension systems receive the most rigorous treatment. This text develops their behavior from first principles, presents design from a physical perspective, and explores both electromagnetic and electrodynamic levitation in-depth. It includes a detailed discussion of electrodynamic systems suitable for very high-speed transportation, including analysis of induced currents, electrodynamic forces, and levitation-drag relationships for various configurations. The treatment of Halbach array-based systems is useful.

The author finishes this text by presenting a history of maglev (magnetic levitation for high speed trains) development from the early twentieth-century conceptual ideas to the technologically sophisticated passenger-carrying test vehicles of today in some countries. Indeed, the Japanese test vehicle at Yamanashi currently holds the world speed record for guided ground transport at 603 km/h; a precursor to Tokyo-Nagoya service at up to 500 km/h, by the mid-2030s. A notable strength of this book is its breadth combined with analytical depth. It covers ships and aircraft, ultrasonic and magnetic levitation, and other systems for which lift forces play a key role. The inclusion of numerical examples enhances its value, helping readers to move from theory to practical systems.

This text fills an important gap in the technical literature in that, to my knowledge, there is no other single volume that offers such a cross-sectional, quantitatively oriented treatment of lift and levitation phenomena across so many domains. It will be helpful for graduate students and engineers in mechatronics and related fields, providing a good broad introduction to the field and a foundation upon which the researcher can build their particular specialty. I am pleased to recommend this text as a reference and a stimulating guide to the field of lift and levitation.

Tony Eastham
Professor Emeritus of Electrical & Computer Engineering
Queen's University, Kingston, ON, Canada

Former Vice-President for Research & Development
Hong Kong University of Science & Technology, Clear Water Bay, Hong Kong
April 2026

Preface

The invention of trains in the early nineteenth century made mass transportation possible, and the subsequent invention of the aircraft brought about a revolutionary means of high-speed transportation unaffected by land terrain. The aircraft brought about a dramatic change in the scale of travel in terms of speed and freedom of movement. Presently, magnetically levitated vehicles traveling above the guideway without contact have continued to be developed to aim for the next phase of mass transportation. Besides, drones that fly freely through the air have attracted much attention.

The magnetically levitated vehicles that generate propulsion and guidance forces without contact can be the alternative to the aircrafts where guideways can be installed. They are less susceptible to weather, eliminate access time to the airports located in suburbs, and would not pose a risk to passengers' lives even if propulsion or thrust forces were lost during travel. Helicopters and drones are capable of vertical takeoff and landing, where drones are characterized by their simpler structure and easier control, leading to a wide range of applications compared to helicopters. Then, updating an overview of the levitation/lift technologies could lead to another breakthrough for future vehicles and equipment.

There are a variety of situations for floating, lifting, or levitating. Objects are suspended in water, air, sound-induced pressure fluctuations, optical beams, or electromagnetic forces. Some objects are made to float for processing in factories; balloons move through the air; ships and hovercrafts cruise on the water; drones, fixed-wing aircrafts, helicopters come and go in the sky; and magnetically levitated vehicles transport passengers at very high speeds with no sound other than the air rushing past it.

The chapter of lifting by fixed wings clarifies the debate about the lift force mechanism of fixed-wing aircraft, which is still misunderstood even in some articles. Drones are now being applied in a variety of fields, and the chapter of rotating wings provides a detailed discussion of their dynamics. Ultrasonic levitation is a unique technique that levitates small objects or relatively heavy planar objects without using a container. The associated chapter introduces analytical methods and demonstrates its usefulness through numerical examples. Detailed and advanced discussions are given for electromagnetic levitation, beginning with determining the controller gains from a physical perspective to construct an electromagnetic levitation system. Regarding the electrodynamic levitation system, which is suitable for super-high-speed transportation, the book shows analytical approaches in detail to solve the advanced problems for eddy currents and electromagnetic forces. The detailed analysis of the electrodynamic levitation system results in some novel findings: the quantitative differences in levitation/drag ratios for various electrodynamic levitation configurations. The novel analytical approach is also given to the electromagnetic field produced by the Halbach array permanent magnets, which is applied to a small-scale electrodynamic levitation vehicle instead of using high-cost superconducting magnets.

Kitakyushu City, Japan Tetsuzo Sakamoto

Competing Interests The author has no competing interests to declare that are relevant to the content of this manuscript.

Contents

About the Author

Tetsuzo Sakamoto is Professor Emeritus of Kyushu Institute of Technology, Japan. His research interests include the electromagnetic analysis of electrodynamic levitation vehicle, the control system design of electromagnetic levitation, the modeling and control system design of linear synchronous motor, and the decentralized control system design and analysis of web tension control system. He has authored two books: (1) T. Sakamoto: "Electromechanical dynamics and Control of Electric Machines," Morikita Pub. Co., Japan (2007), which received the best book award from The Japan Society of Applied Electromagnetics and Mechanics (2) T. Sakamoto: "Fundamentals and Control of Synchronous Motor," Morikita Pub. Co., Japan (2023).

Chapter 1
Buoyancy

Abstract As an introduction to the concept of lift/levitation forces, the chapter discusses the mathematical relationship between atmospheric pressure and density with respect to altitude. Then, in this chapter the physical phenomena of buoyancy acting on objects in water and balloons floating in the atmosphere are described in details.

1.1 Pool and Atmosphere

The upward force acted on a body completely or partially submerged in water at rest is widely known as Archimedes' principle. For example, imagine you enter a pool that is deep enough that your feet can just barely touch the bottom as shown in Fig. 1.1. You will then feel as if your body weight is lighter, and you will be able to move back and forth and side to side with ease. This is because your body receives a force equal to the weight of the water it expels, which is called buoyancy.

Incidentally, the air density ρ_a is much smaller than that of water, and its density at altitude $h = 0$ on Earth is $\rho_a \mid_{h=0} = \rho_{a0} \cong 1.2\,\text{kg/m}^3$, which is about 1/800 of that of water. In other words, 1 cubic meter of air has a mass of about 1.2 kg, which is almost the same as a laptop computer. Ninety percent of the atmosphere exits below the altitude of about 16 km, and the atmospheric pressure is due to the weight of the air above. If the air density were constant regardless of altitude, the atmospheric pressure p_a would decrease at a constant rate as the altitude h increases. However, according to Boyle's law, the density ρ_a varies in proportion to the pressure p_a at a constant temperature, and in the troposphere, temperature decreases at a constant rate. If we use $\gamma = 6.5 \times 10^{-3}$ K/m as the coefficient for altitude h[m], the absolute temperature is given as

$$T(h) = T_0 - \gamma h \tag{1.1}$$

T. Sakamoto, *Fundamentals of Producing Lift Forces*,
https://doi.org/10.1007/978-981-92-2222-3_1

Fig. 1.1 Buoyancy

where $h \leq T_0/\gamma = h_\infty$. The differential equation for the pressure in terms of altitude is given by

$$dp_a(h) = -\rho_a(h)gdh \tag{1.2}$$

where g is the gravitational acceleration. And, the pressure has the relation with the density expressed by

$$p_a(h) = \frac{RT(h)}{M}\rho_a(h) \tag{1.3}$$

where R is the gas constant, and M the molar mass. Then, the solution p_a [hPa] of Eq. (1.2) is given by the following equation:

$$\begin{aligned} p_a(h) &= p_{a0}\left\{\frac{T(h)}{T_0}\right\}^{g/\gamma R_M} \\ &= 2.20 \times 10^{-6} \times (44.3 - 0.001h)^{5.25} \end{aligned} \tag{1.4}$$

where p_{a0} and T_0 are the air pressure and the absolute temperature on the ground, respectively. The density ρ_a [kg/m^3] is then given by the equation

$$\begin{aligned} \rho_a(h) &= \rho_{a0}\left\{\frac{T(h)}{T_0}\right\}^{g/\gamma R_M - 1} \\ &= 1.233 \times 10^{-7} \times (44.3 - 0.001h)^{4.25} \end{aligned} \tag{1.5}$$

where $g = 9.80\,\text{m/s}^2$, $\gamma = 0.0065\,\text{K/m}$ the constant of temperature drop to altitude increase, and $R_M = R/M = 287\,\text{J/K kg}$. Figure 1.2 shows the calculated results of Eqs. (1.4) and (1.5). The mass of air pillar with the area $1\,\text{cm}^2$ on the ground corresponds to the atmospheric pressure value shown in Fig. 1.2. The exact atmospheric pressure on the ground is 1 atm = 1013 hPa = 1.034 kg f/cm^2, while the mass of the above-mentioned air pillar with the area of $S_p = 1\,\text{cm}^2$ is derived as

$$\begin{aligned} m_a(0) &= S_p\int_0^{h_\infty}\rho_a(h)dh = S_p\frac{p_{a0}}{g} \\ &= 10^{-4} \times \frac{1.013 \times 10^5}{9.80} = 1.034\,\text{kg} \end{aligned} \tag{1.6}$$

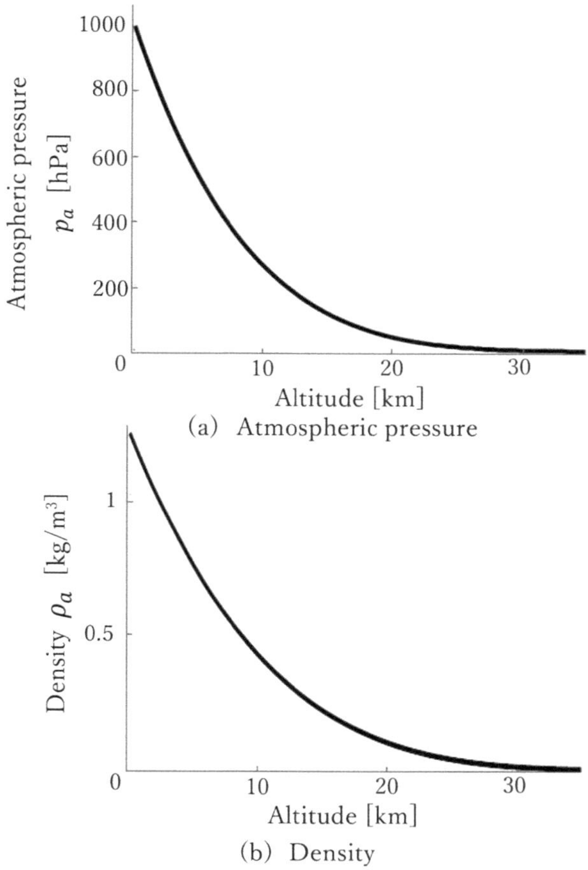

Fig. 1.2 Atmospheric pressure and density vs. altitude

If a mass of warmer air appears in the air, the density is smaller than the surrounding area by the temperature, so the buoyancy in the atmosphere causes the warmer air to rise. In other words, the buoyancy that occurs in a pool is the same as the phenomenon in the atmosphere.

1.2 Floating Object in Water

As shown in Fig. 1.3, a cuboid with a height H [m], a bottom area S [m^2], and a density ρ [kg/m^3] is placed in a tank containing water. Now let's consider the equation for buoyancy, assuming that the object is floating h [m] above the water surface, when the density of water is ρ_w [kg/m^3] and the atmospheric pressure is p_a [N/m^2].

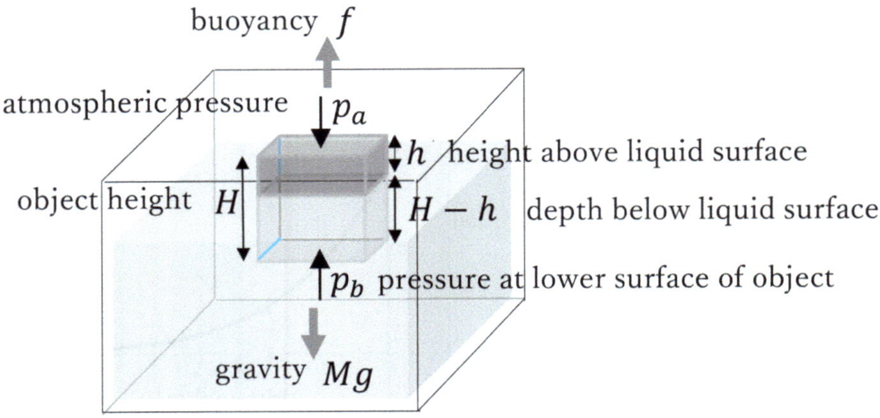

Fig. 1.3 Buoyancy on object

The atmospheric pressure p_a acts on the top surface of the object, while the pressure p_b created by water acts on the bottom surface, which is greater than atmospheric pressure by the depth $H - h$ [m] from the water surface. The pressure p_b [N/m^2] is given by

$$p_b = p_a + \rho_w (H - h) g \tag{1.7}$$

And, the buoyancy f [N] is produced by the pressure difference between the top and bottom surfaces:

$$\begin{aligned} f &= p_b S - p_a S \\ &= \rho_w (H - h) S g \end{aligned} \tag{1.8}$$

This equation shows that the object weight becomes smaller by the water volume $(H - h)S$ [m^3] expelled by the object.

On the other hand, the gravity acting on the object is $Mg = \rho H S g$ [N], where $M = \rho H S$ [kg] is the mass of the object. Gravity acts downward and buoyancy acts upward on the object, so when these two forces are equal, balance is achieved. The positive height h of the object above the water surface represents that the object floats due to the buoyancy. Then, the equation $f = Mg$ gives the equilibrium relation:

$$\rho_w (H - h) = \rho H \tag{1.9}$$

If the density of the object ρ is less than the water density ρ_w, then $h > 0$, i.e., the object is floated in water (Fig. 1.4).

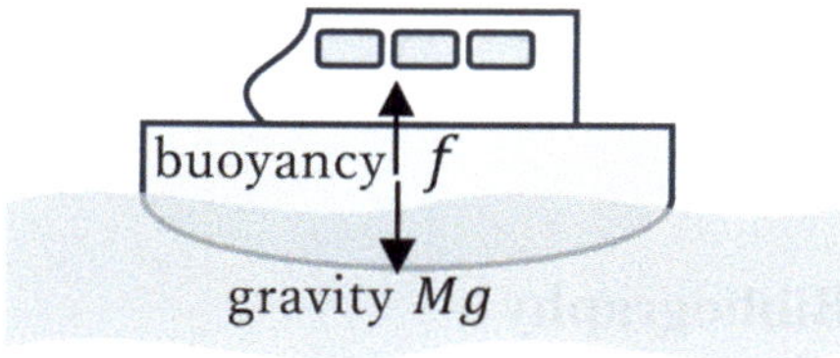

Fig. 1.4 Buoyancy and gravity

1.3 Balloon

Balloons float in the same way as ships, following Archimedes' principle. If the average density of a balloon is less than the density of the air surrounding it, it can obtain buoyancy. The gas inside a balloon is not limited to air, but let's think of air as the gas inside the balloon in Fig. 1.5. The buoyancy f of a balloon is determined by $\rho_a Vg$, where V is the volume of the balloon displacing the air, and ρ_a is the surrounding air density, which changes with air pressure and temperature. Meanwhile, the density of the gas in the balloon ρ_h also changes with temperature, so gravity Mg can be controlled by changing the temperature of the gas inside the balloon. Here, the mass M is made up of the gas part $M_h = \rho_h V$ and the remaining part M_c, and can be written as $M = \rho_h V + M_c$. Thus, the equation for buoyancy and gravity is

$$\rho_a Vg = \rho_h Vg + M_c g \tag{1.10}$$

where the left-hand side (lhs) of the equation is the buoyancy, and the rhs is the gravity. As the temperature is increased the balloon gas density ρ_h decreases. On the other hand the atmosphere pressure decreases as the balloon goes up. Therefore

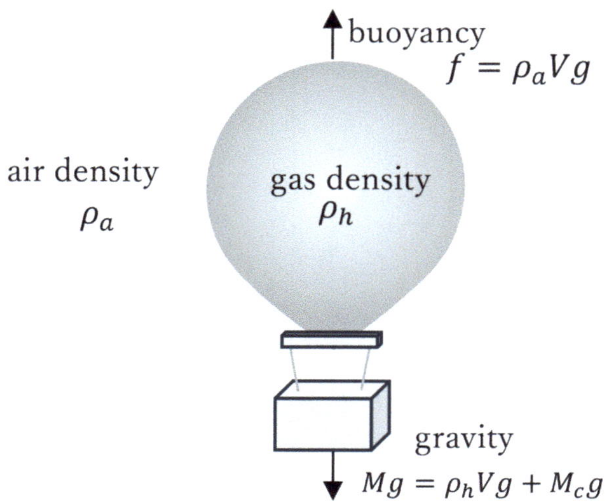

Fig. 1.5 Balloon

temperature rise of balloon makes it rise higher. Incidentally, it is well known that helium is used as the balloon gas.

Bibliography

1. Y. Ogura, *General Meteorology* (in Japanese) (The University of Tokyo Press, 2016)

Chapter 2
Lift Force Produced by Fixed Wing

Abstract To understand the lift generated by fixed wings of an aircraft, three factors are necessary to understand: the Coanda effect, the angle of attack of the wing, and the law of action and reaction. The chapter provides a physical explanation based on the book "Understanding FLIGHT" by Anderson and Eberhardt, published in 2000, which revised the previous interpretation based on "Equal Transit Time theory."

There are two types of wings: fixed and rotary; the former are fixed to the aircraft body, while the latter generate lift by rotating, as seen in helicopters and drones. Fixed-wing aircraft is characterized by high-performance flight, while rotary-wing aircraft has the advantage of having a high degree of freedom of movement and the ability to hover at low speeds or even stationary.

2.1 Lift Force

Figure 2.1 shows ideal streamlines with no separation of the air flow on the surface when the wing is moving to the left of the figure. The floating force described in Chap. 1 was called buoyancy, but the floating force created by the fixed wings of an airplane is called lift force. The book "Understanding FLIGHT" authored by D. Anderson and S. Eberhardt, which was published in 2000, contained groundbreaking content that overturned conventional wisdom regarding the explanation of how lift is generated. The concept of the mechanism of lift generation had to be completely revised, where the main part of the book is available online. The description here basically follows this book. The previous interpreting for lift force generation is discussed in Sect. 2.2.

Figure 2.1 illustrates the air flow around a fixed wing. When a wing moves at a constant speed to the left of the figure, the air that hits it creates the highest pressure at the leading edge of the wing, creating a stagnation point where the flow velocity is zero. The air moving toward the upper surface of the wing is accelerated by the

T. Sakamoto, *Fundamentals of Producing Lift Forces*,
https://doi.org/10.1007/978-981-92-2222-3_2

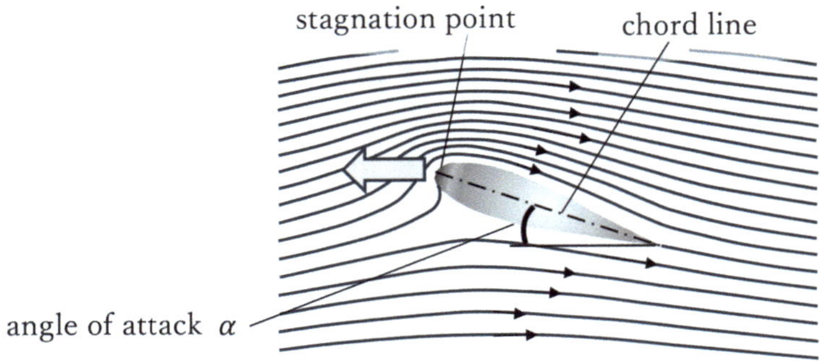

Fig. 2.1 Airflow over a fixed wing

high pressure generated at the leading edge, and therefore the pressure decreases in opposition to the increase in velocity. However, after reaching a maximum velocity near the maximum wing thickness, the flow slows down, while the pressure on the underside of the wing increases as the wing acts as an obstacle for the flow. However, since air has viscosity, friction on the wing surface creates losses, so that the acceleration on the upper surface of the wing gets smaller due to the losses compared with a non-viscous fluid. Accordingly the flow on the upper surface of the wing arrives at the trailing edge of the wing earlier than the underside. A supplementary explanation of this will be given later.

The mechanism of lift generation can be summarized as the result of the three factors described below.

2.1.1 Air Flow Influenced by Coanda Effect

When a curved surface is present in the flow of a fluid, the changes in pressure and speed of the fluid, as well as the effects of viscosity, cause the flow to be more attracted to the curved surface. The flow of water coming out of a tap draws in air molecules from nearby areas, but then the lost air is replenished from the surrounding area. If a cup is placed in the flow of water as shown in Fig. 2.2a, the small amount of air between the water and the surface of the cup is drawn into the flow of water. However, since air molecules cannot be replenished in this case, the pressure on the surface of the cup decreases by that amount. Also, the water viscosity makes the force of drawing the water to the cup surface larger. As a result, the water flow is attracted to the surface of the cup, and its direction is altered after it passes through the cup. And as a reaction to this force, the reverse force acts on the cup. This is called the Coanda effect.

Figure 2.2b shows the Coanda effect occurring on the upper surface of a wing. Airflow with a speed difference relative to the wing draws in air between the flow and the wing, reducing the pressure on the wing surface. Moreover the air viscosity

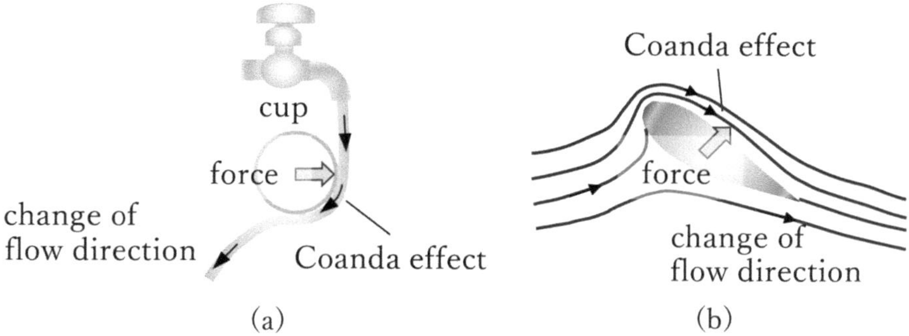

Fig. 2.2 Coanda effect. (**a**) Coanda effect of water. (**b**) Coanda effect of air

draws in more air strongly. At this time, the greater the airspeed, the greater the pressure drop, and thus the more airflow can be maintained. On the underside of the wing, the airflow suddenly decelerates and the pressure becomes higher.Then, the greater the drop in pressure on the upper surface of the wing compared to the underside, the greater the lift, which shows that the flow is asymmetric between the top and bottom surfaces as shown in the figure. The figure also shows that the airflow that has passed the wing changes its direction downwards. Wings come in a variety of cross-sectional shapes, but giving them a smoothly curved surface provides greater lift with the Coanda effect.

2.1.2 Downward Airflow Caused by Angle of Attack

The straight line connecting the leading and trailing edges of a wing is called the chord line (see Fig. 2.1), the angle between this line and the direction of flight the angle of attack, and the length of the chord line the chord length. Increasing the angle of attack α increases the downward flow of air behind the wing, so that lift increases proportionally within a certain range of angles of attack. However, if the angle of attack is increased beyond a certain value, the air flow will separate from the wing surface, i.e., the Coanda effect will be lost. This will result in a loss of lift, and in the case of an airplane it will stall.

Even with a flat, plate-like wing, lift is generated if there is an angle of attack, but the Coanda effect that guides the airflow is weak. Accordingly, the wing has a poor value of lift to drag ratio, in which the value is called the lift-to-drag ratio (Fig. 2.3).

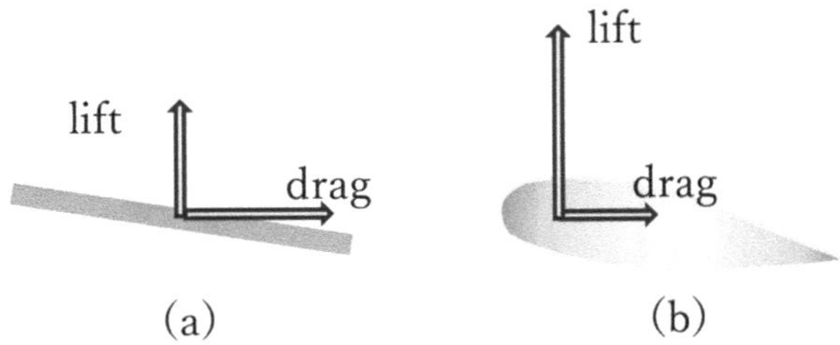

Fig. 2.3 Lift/drag ratio difference depending on fixed-wing cross-section. (**a**) Plate-shaped wing. (**b**) Wing with curved surface

2.1.3 Lift Produced by the Law of Action and Reaction

Imagine the air flow around the front and rear of a wing (see Fig. 2.1), the flow is redirected downward at the rear compared to the front. The flow redirection produces lift on the wing, which translates into the law of action and reaction. In other words, instead of being interpreted as being due to a pressure difference as mentioned above, it's possible to understand that the lift is generated by the result of changing the momentum of the air flow.

The following three basic quantitative relationships hold for lift generation.

- The mass of downward air component per unit time and cross-sectional area that is created by the wing m_{div} is proportional to the wing's cruising speed v and air density ρ_a:

$$m_{div} \propto v\rho_a \text{ [kg/m}^2\text{ s]} \tag{2.1}$$

- The downward component of air velocity created by the wing v_{ver} is proportional to the wing's cruising speed v and the angle of attack α:

$$v_{ver} \propto v\alpha \text{ [m/s]} \tag{2.2}$$

- The lift force F_L is proportional to the product of the mass of air m_{div}, which is deflected downward by the wing, and the downward air velocity v_{ver}:

$$F_L \propto m_{div} v_{ver} \text{ [kg/ms}^2\text{]} \tag{2.3}$$

From these relationships, and with the use of the lift coefficient C_L that includes the effect of the angle of attack α, the lift F_L is expressed by the following equation:

$$F_L = \frac{\rho_a S}{2} C_L v^2 \text{ [N]} \tag{2.4}$$

where S [m^2] is the surface area of the wing. The lift is proportional to the lift coefficient C_L, which is proportional to the angle of attack α within a certain range, the air density ρ_a, and the wing area S, and is also proportional to the square of the wing's cruising speed v. In terms of physical quantities that an airplane pilot can control, lift F_L is controlled by changing the angle of attack α and the wing's cruising speed v. Therefore, in order to fly an airplane against the gravity, the angle of attack is increased immediately after takeoff due to the low speed, and then as the speed increases sufficiently the angle of attack is decreased.

2.2 Problem with the Previous Understanding of Lift Mechanism

In December 2000, D.F.Anderson and S.Eberhardt published the book "Understanding Flight," which pointed out four problems with the explanation of lift generation that had been generally accepted around the world until then. Figure 2.4 is one of the most representative diagrams.

First, the diagram makes the assumption that the air flowing from the leading edge of the wing takes the same amount of time to reach the trailing edge at the top and bottom of the wing. This is called the Equal Transit Time theory, and subsequent research has shown that this is incorrect, as the air at the top arrives earlier than the bottom, as mentioned above. Despite there being no evidence that they arrive at the same time, this was the commonly held belief for a long time.

The second point is that the typical cross-sectional shape of a wing used in the explanation is straight on the bottom, but rounded on the top, as shown in the figure. In other words, the path of the air flow is longer at the top and shorter at the bottom, and the Equal Transit Time theory is applied to this. If the lengths of the paths are different, then the speed at the top will be faster by the length of the path, it says. However, it is known that calculating lift from the speed difference of the air flow is only an insufficient amount. Furthermore, according to this explanation, it would be impossible for an airplane to fly upside down. In other words, it cannot explain the lift generated by inverted flying.

The third point is that in this diagram, the air flow is horizontal both in front of and behind the wing, in other words the air flow after the trailing edge of the wing is in the same direction as the air flow in front of the wing. According to Newton's law of action and reaction, the air flow behind the wing should have a downward component equal to the amount of lift generated. There is no evidence in the streamlines that the wing is floating due to lift, and thus there is a clear error in the description of the air flow.

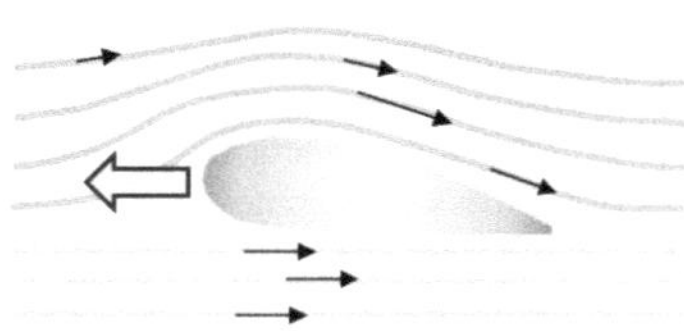

Fig. 2.4 Typical example of wrong depiction

The fourth point is the application of Bernoulli's theorem to the velocity difference between the upper and lower surfaces of a wing under the theory of Equal Transit Time. The theorem states that for a fluid flow with no viscosity and constant density, where density is ρ, velocity is v, and pressure is p, the sum of the dynamic pressure $p_d = \rho v^2/2$, which corresponds to the kinetic energy of the fluid, and the static pressure $p_s = p$, which corresponds to the potential energy of the fluid, is a constant value on the streamline:

$$p_d + p_s = \frac{1}{2}\rho v^2 + p = \text{const.} \tag{2.5}$$

where the dynamic pressure p_d is created by density ρ and velocity v, while the static pressure p_s by a fluid in a stationary state. Thus, it can be seen that the lift equation (2.4) is expressed as the product of the dynamic pressure p_d and the wing area S. That says, the high velocity v on the upper surface produces lower pressure p according to Eq. (2.5), while the low velocity v creates a larger pressure p, generating lift force. However, Bernoulli's theorem applies to the steady flow of non-viscous, incompressible fluids. Air has both viscosity and compressibility, and there is a boundary layer where the velocity on the wing surface changes from 0 to a uniform flow velocity due to viscous friction. Hence, it is not technically appropriate to apply Bernoulli's theorem in this case. It is said that the quantitative relationship between the velocity v and the pressure p expressed in Eq. (2.5) does not hold true accurately even if the boundary layer is excluded from the calculation.

To summarize, it is important to note that the Equal Transit Time theory used in previous interpretations is not valid. As the theory overly underestimates the speed difference between the top and bottom of the wing, the calculated lift results in too small value. Another point is that Bernoulli's theorem cannot be applied to the airflow in a strict sense. Additionally, there are cases where the representation of the air streamlines in front of and behind the wing does not have any physical validity. This poses a major problem in determining that lift is being generated from the shape of the streamlines, and there is also a problem with the shape of the wing used in the explanations.

Bibliography

1. Cambridge University, Airflow across a wing. https://www.youtube.com/watch?v=UqBmdZ-BNig
2. D. Anderson, S. Eberhardt, How airplanes fly: a physical description of lift. The Aviation History Online Museum (2015)
3. T. Hasegawa, The plane is flying like this! https://h-a-labo.amebaownd.com/posts/6415375
4. T. Benson, Incorrect theory #1. https://www.grc.nasa.gov/WWW/k-12/VirtualAero/BottleRocket/airplane

Chapter 3
Force Produced by Rotating Wing

Abstract The chapter provides a detailed explanation of the physical phenomena involved in helicopter flight. First, it describes the required cross-sectional shape of the blades, as well as the types, structures, and control of the hinges connected to the hub. Formulations are given to explain the physical phenomena, including the precession essential to helicopter dynamics with rotor blades, and the mechanical problems caused by the non-uniform relative wind speed depending on the rotational position of the rotor blades. Furthermore, the chapter provides a detailed discussion of drone dynamics, which have seen active application in recent years, using mathematical formulas not only for lift but also for the motion of the body.

3.1 Bamboo-Copter

Figure 3.1 shows a bamboo-copter; by partially scraping the wing, air flows are created in the same way as with airplane wings, generating lift. When it is rotated as shown in figure (a), the part of the wing that is scraped creates a downward air flow as shown in the figure (b), generating lift.

3.2 Airplane Propeller

An airplane propeller is shown in Fig. 3.2. It is a rotary wing that uses the lift generated in the case of a fixed wing as thrust in the forward and backward directions. In a rotary wing, as the peripheral speed increases in proportion to the distance from the center. And, as can be seen from the lift equation of Chap. 2 that indicates that the lift is proportional to speed, the thrust increases in proportion to the square of radius. Therefore, if the local angle of attack over the length of the blade, called the pitch angle, is varied inversely proportional to the position from the center, then the stress distribution on the blade will be proportional to the distance from the center. Furthermore, it is possible to make a uniform stress distribution along the blade length by devising a cross-sectional shape. Accordingly, as shown in the figure the blade has a twisted shape with making thicker to get stronger towards the

T. Sakamoto, *Fundamentals of Producing Lift Forces*,
https://doi.org/10.1007/978-981-92-2222-3_3

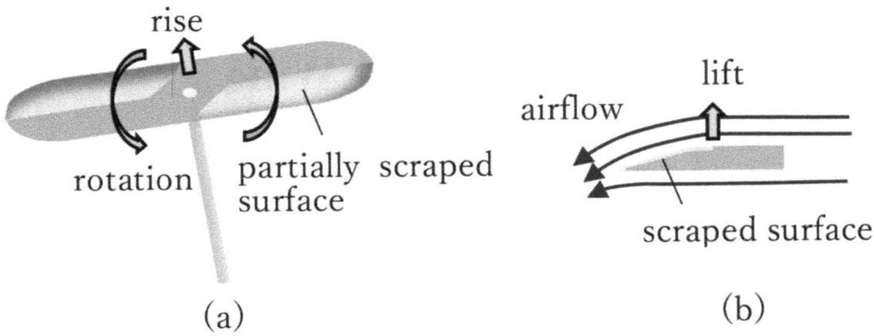

Fig. 3.1 Bamboo-copter. (**a**) Exterior. (**b**) Airflow

Fig. 3.2 Twist of propeller

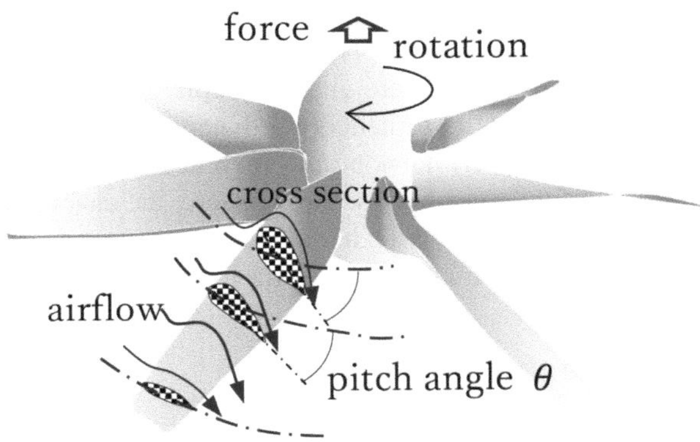

center, while making thinner on the outside. Therefore, especially the cross section of the central part of the blade has a shape similar to that of a fixed wing. Also, if the output of the engine that drives the propeller is increased, the generated force can be increased, in which the number of blades is increased, and/or the blade length is increased.

3.3 Hovercraft

When an airplane approaches the runway and its altitude drops to less than half the width of the wing, the air between the wings and the ground becomes compressed, generating larger lift relative to the speed, facilitating a smooth landing. This is called the ground effect. Therefore, by utilizing this effect, even airplanes with small thrust can fly with sufficient lift at a very low altitude.

The name "hovercraft" is widely known, but this was originally a trademark name. The official name is air cushion vehicle (ACV), but in this book we use the former, which is now widely used as the de facto generic name. Hovercrafts employ the ground effect to form an air cushion with the air jet created by the

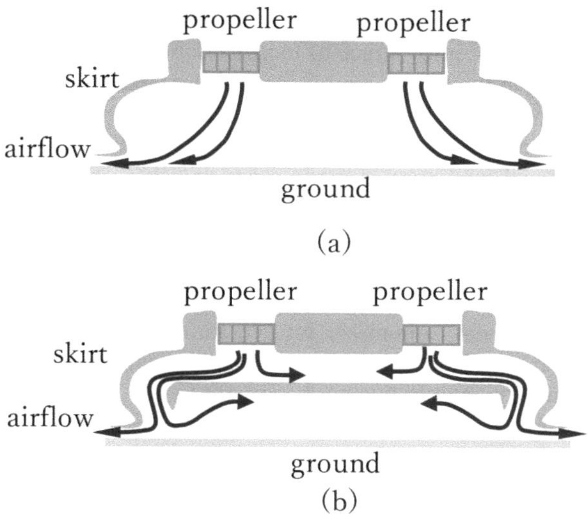

Fig. 3.3 Structure of hovercraft. (**a**) Plenum chamber craft. (**b**) Peripheral jet craft

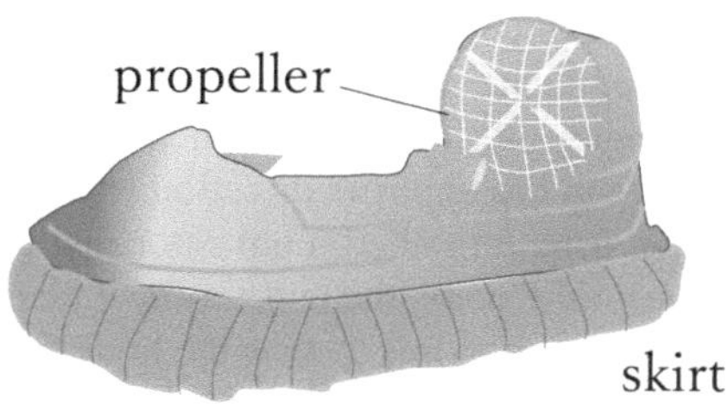

Fig. 3.4 Hovercraft

propeller, allowing them to move floating on land or water. Figure 3.3 shows two types of hovercraft designed for land use. The closed chamber type shown in figure (a) creates an air jet with the propeller, and when it gets past obstacles on land or water waves, a synthetic rubber part called a skirt plays the role of maintaining the pressure of the cushion as it deforms. This type uses the ground effect to float, but has the advantage of being a simple structure. Figure (b) shows that when a jet of air is created, a small cross-sectional jet is created only along the periphery by the components including the skirt. In this type, the ground effect inside the skirt is maximized, generating a large lift. This is called a peripheral jet, and it is possible to form a cushion of high-pressure air on the underside of the body. Figure 3.4 shows an overview of a small type of hovercraft.

3.4 Helicopter

In helicopters, which float using the airflow created by the rotors, the speed and mass of the air blown down determines the lift, and the greater the air speed, the greater the impact on the ground. Therefore, to minimize this impact, it is desirable to be able to blow as much air down as possible. However, if the length of the rotor blades becomes larger to increase the amount of air blown down, the centrifugal force at the tips of the blades increases, leading to the possibility of blade rigidity problems. Furthermore, the problem of vibration associated with blade rotation becomes more pronounced. Helicopters are relatively familiar aircraft that can take off and land even in urban areas, but they are also technically very complex and intriguing aircrafts in terms of precession, as will be described later.

Helicopter engines are usually turboshaft engines consisting of a compressor, combustion chamber, and turbine, but inexpensive reciprocating engines also seem to be used for small helicopters. The rotor, consisting of multiple rotating blades, is the element responsible for generating the main force and thus the main actuator of the helicopter. As an actuator, the rotor generates force roughly in the axial direction, and in this book we refer to this force as thrust force since it is the force generated by the actuator. Thrust is divided into a lift force component for lift and a propulsion force component in the direction of travel (Fig. 3.5). In addition, the rotor rotation direction can be commercially categorized into counterclockwise (CCW) and clockwise (CW) models when viewed from above. In this book we take the model of CCW rotation type.

A rotorcraft with one main rotor is called the single-rotor helicopter (Fig. 3.6), while the one with two main rotors is called the twin-rotor helicopter. In the case of a single-rotor helicopter, when torque is applied to the main rotor, the aircraft will rotate in the opposite direction to the main rotor. Therefore, as shown in the figure, if an actuator called a tail rotor that generates an auxiliary force is attached to the rear of the body, the anti-torque T_a can be applied to the body as an external force, making it possible to stop the body from rotating.

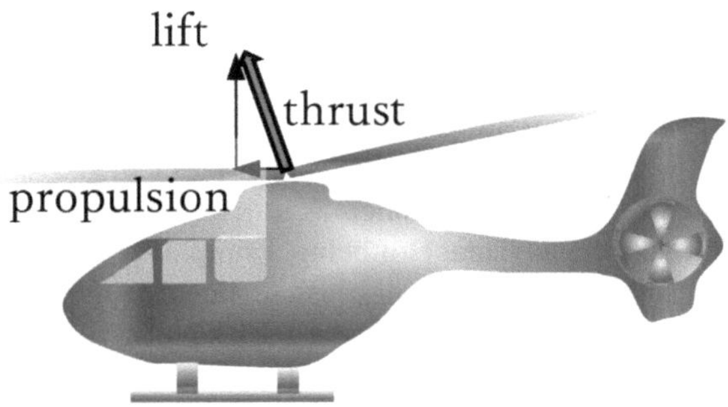

Fig. 3.5 Thrust dividing to lift and propulsion

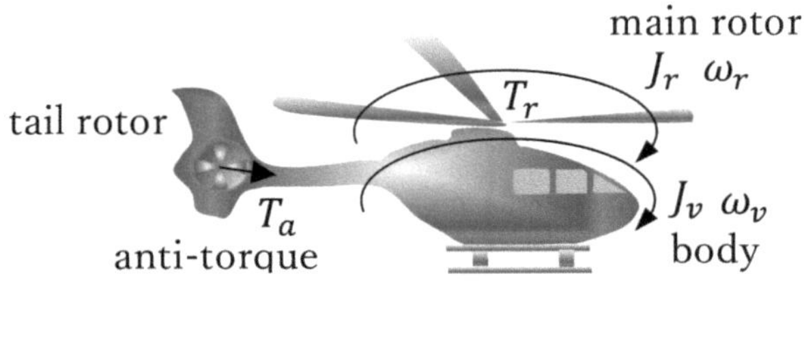

Fig. 3.6 Single-rotor helicopter

Let's consider some simple mechanical formulations to confirm the situation in which the body's rotation is prevented in the case of a single-rotor aircraft. In Fig. 3.6, if the main rotor rotates, the helicopter body rotates with the theorem of conservation of angular momentum. Then, assuming the anti-torque T_a[Nm] of the tail rotor, we obtain

$$T_a = J_r \frac{d\omega_r}{dt} + J_v \frac{d\omega_v}{dt} \tag{3.1}$$

where J_r [kg m^2] and J_v [kg m^2] are the moments of inertia for the main rotor and the body, respectively, and ω_r [rad/s] and ω_v [rad/s] are the angular velocities for the main rotor and the body, respectively. Moreover, defining the main rotor torque as

$$T_r = J_r \frac{d\omega_r}{dt} \tag{3.2}$$

we get the helicopter body acceleration

$$\frac{d\omega_v}{dt} = \frac{T_a - T_r}{J_v} \tag{3.3}$$

This relation shows that by adjusting the anti-torque T_a the helicopter's rotational angular velocity ω_v can be controlled to 0. In the case of the twin-rotor aircraft, the angular momentum can be cancelled out by reversing the rotation direction of both rotors, but there are some ways to arrange the rotors. One is the counter-rotating coaxial rotor aircraft, in which two rotors are installed on the same axis, and the other is the tandem-rotor aircraft, in which rotors are installed at the front and rear of the aircraft and rotate in opposite directions.

3.4.1 Rotor Blade

The blades are basically the same shape as those of propeller aircraft, but in the case of airplanes, the propellers rotate perpendicular to the ground during takeoff, so that longer blades impose limitations on ground clearance. However, for helicopters, the propellers rotate parallel to the ground, and thus such limitations are weaker, and longer blades are even more efficient in terms of generating lift, since they can cause a large change in the momentum of the air. However, if the blades are made too long, there is a problem of mechanical instability, as the peripheral speed of the blade tips approaches the speed of sound. And while the number of blades increases the lift increases. However, the mechanical structure becomes complex, and the weight also increases, causing the "flapping" sound that is unique to helicopters in addition. In other words, the air blown down is a phenomenon in which a large amount of air is knocked down by the blades, which leads to the loudness of the

Fig. 3.7 Rotor blade

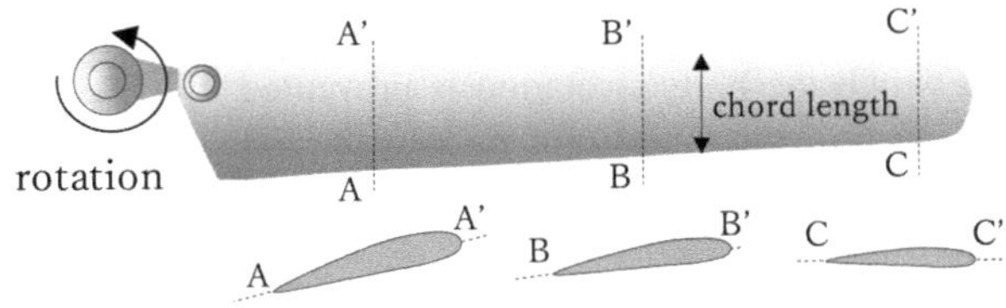

Fig. 3.8 Rotor structure

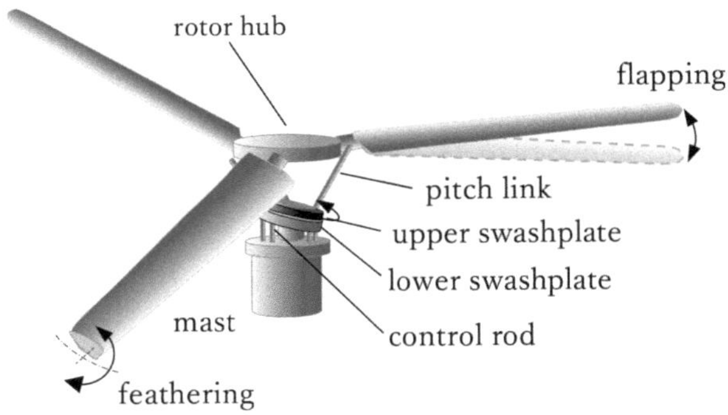

sound generated. Also, in fixed-wing aircraft, thrust is produced by the propeller or jet engine, and lift is produced by the wings. But in helicopters, the thrust generated almost perpendicular to the rotational plane of the main rotor is allocated to both lift and thrust, in which the complexity exists.

Figure 3.7 shows an example of a blade shape. The airspeed of a rotating blade increases as the distance increases outward from the center of rotation. If the blade has the same cross-sectional shape, the outer side will then generate greater lift, so some method is required to average it out. Lift is proportional to both the pitch angle and the chord length. Therefore, in order to average the internal stress and lift along the entire length of the blade, the blade can be tapered by flattening the cross-section toward the outer side and shortening the chord length, or the blade can have a twisted cross-sectional shape along its length as shown in the figure. Therefore, the pitch angle is made larger at the base and smaller toward the tip. The planar shape of a blade viewed from above is often a rectangular in practice, because of easy manufacturing.

3.4.2 Rotor Structure

As shown in Fig. 3.8, the rotor blades are held in place by joint parts in a part called the rotor hub, and there are three basic joint parts connecting the hub and the blades: the flapping hinge, the drag hinge, and the feathering hinge. The flapping hinge enables the blades to rotate in the up and down directions shown in the figure, which is called flapping. This plays a role in mitigating the up and down bending stress acting on the connection parts, which is caused by the lift force acting on the blades, gravity, and the centrifugal force associated with rotation.

The feathering hinge is used to control the pitch angle, where the swashplate plays an important role. The swashplate consists of an upper swashplate that rotates with the rotor, and a lower swashplate that moves only up and down by the control rod. For example, when the swashplate rises, the pitch link in the diagram rises, and the blades rotate around their length axes according to the amount of movement, increasing the pitch angle. Therefore, if all the control rods are moved the same amount, the entire swashplate moves up and down in parallel, and the pitch angles of all the blades are controlled to the same amount. This is called the collective pitch control. If the length of the control rod in the diagram is partially changed, the swashplate will tilt diagonally as shown in the diagram. Then the pitch angle of the blades will change at a specific orientation when viewed from the aircraft. This is called the cyclic pitch control, and in this case, lift will increase or decrease in a specific direction.

A joint called a drag hinge relieves concentrated loads along the rotational surface acting on the hub connection.

Regarding these three basic hinges, there are types that have only two of the flapping hinges, the feathering hinge, and types that have only the feathering hinge. Flapping and dragging are intended to relieve concentrated loads on the hub connection, but there are types that relieve concentrated loads by deflecting the blades instead of using hinges. In addition, for flapping, the aerodynamic force acting on the blades acts as a damping effect to suppress vibration.

Figure 3.9 shows the blades before and during takeoff. Before takeoff, only the centrifugal force due to the rotation of the blades is generated, and there is no flapping, i.e. the flapping angle is 0. The blades coincide with the plane connecting the blade tips, which is called the tip path plane. If the engine output is increased and the pitch angle of all the blades is increased by controlling the collective pitch, lift is generated on the blades, and the aircraft's gravity acts on the hub. Then, the blades flap, and the blades have a flapping angle as shown in the figure. The angle of the blade tips relative to the tip path plane is called the coning angle. In addition, in the case of an aircraft type that does not have a flapping hinge, the blade's elasticity causes it to bend in the longitudinal direction.

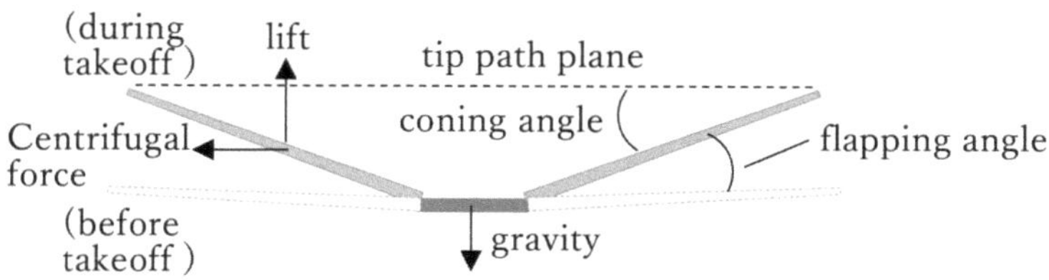

Fig. 3.9 Coning of blade

3.4.3 Rotor Blade Control

As shown in Fig. 3.5, in order to allocate the thrust generated by the rotor not only to lift but also to propulsive force, it is necessary to tilt the rotor axis in the direction of travel. It is readily understood that controlling the cyclic pitch can increase lift in a specific direction, but this is a problem specific to rotating bodies that requires careful consideration of mechanics. Helicopter cockpits are equipped with a cyclic pitch stick, and tilting it forward will cause the aircraft to lean forward, generating thrust and moving forward, but it should be noted that this does not create a difference in lift in the front to back direction. In models where the rotor rotation direction is CCW when viewed from above, increasing lift to the left of the aircraft will cause the rotor's rotation plane to tilt forward, and this is due to the nature of precession, which is caused by the rotor's angular momentum. First, we will explain this in detail.

Figure 3.10 shows a top with its central axis tilted at an angle of θ and rotating at an angular velocity ω. It is well known that the top does not move in the direction of gravity Mg acting on the center of gravity G, but instead precesses. In other words, when an external force acts on a top with some angular momentum, it never falls in the direction of the external force, but continues to keep its motion by moving in a direction different to the external force. Now, let us explain how to express the moment of force (hereafter referred to as torque) and angular velocity as vectors in space. As shown on the right side of the figure, the direction of rotation corresponds to the direction of turning a right-hand thread, and an arrow is drawn in the direction in which the right-hand thread advances, where the arrow is expressed as the torque vector, the angular velocity vector, etc.

If the position vector from the point o where the axis of the top touches the ground to the center of gravity G is $\boldsymbol{r}$, and the unit vector of the z-axis taken perpendicular to the ground from the point o is $\boldsymbol{e}_z$, the torque $\boldsymbol{T}$ [Nm] acting on the top due to the gravity Mg [N] is given by the outer product of the vectors (see Fig. 3.11).

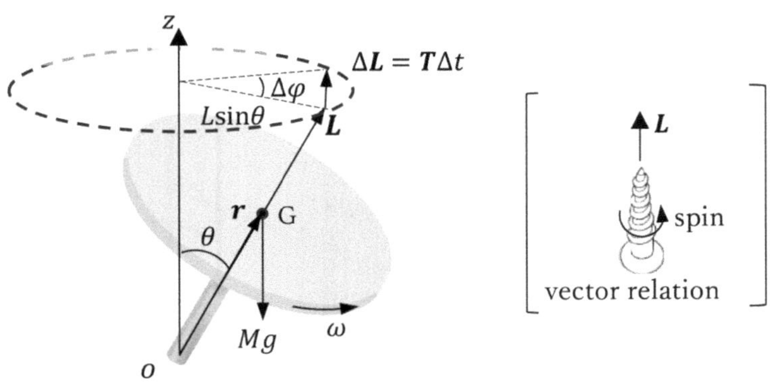

Fig. 3.10 Precession of spinning top

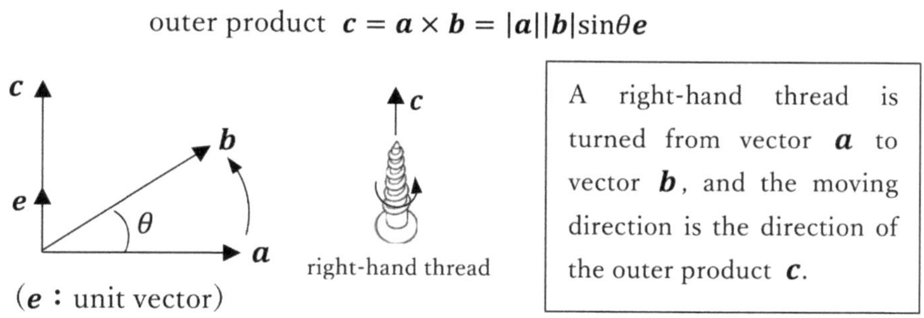

Fig. 3.11 Outer product

$$\boldsymbol{T} = \boldsymbol{r} \times (-Mg\boldsymbol{e}_z) = rMg\sin(\pi - \theta)\boldsymbol{e}_\varphi \tag{3.4}$$

where $\boldsymbol{e}_\varphi$ is the unit vector of the right-hand thread rotation direction about the z-axis. Then we obtain the torque equation

$$T = rMg\sin\theta \tag{3.5}$$

The top has the angular momentum rotating with angular velocity ω, but the gravity-related torque T additionally induces another angular momentum.

Now, if we consider linear motion acting on an object of mass m [kg] with a speed v [m/s] and a force f [N], recall that the change Δp in momentum $p = mv$ [kg m/s] over time Δt [s] is equal to the impulse $f\Delta t$ [N s], being expressed as $\Delta p = f\Delta t$. The torque T caused by the top tilting changes the angular momentum vector $\boldsymbol{L}$ of the top:

$$\boldsymbol{T} = \frac{d\boldsymbol{L}}{dt} \tag{3.6}$$

And, we obtain the change in angular momentum over time Δt as

$$\Delta\boldsymbol{L} = \boldsymbol{T}\Delta t \tag{3.7}$$

As shown in Fig. 3.10, the angular momentum vector $\boldsymbol{L}$ at a position $L\sin\theta$ [Nm s] from the z-axis moves to the position $\boldsymbol{L} + \Delta\boldsymbol{L}$, which indicates that the axis pointing to this position is the new rotation axis after Δt [s].

We have looked at CCW rotation above, but in the case of CW rotation, the angular momentum vector $\boldsymbol{L}$ points in the opposite direction to CCW rotation, as shown in Fig. 3.12, and the external force (gravity) also causes the axis of rotation to tilt in a direction that is $\pi/2$ [rad] behind in terms of rotation. The precession motion is in the same direction as the direction of rotation in either case of rotation.

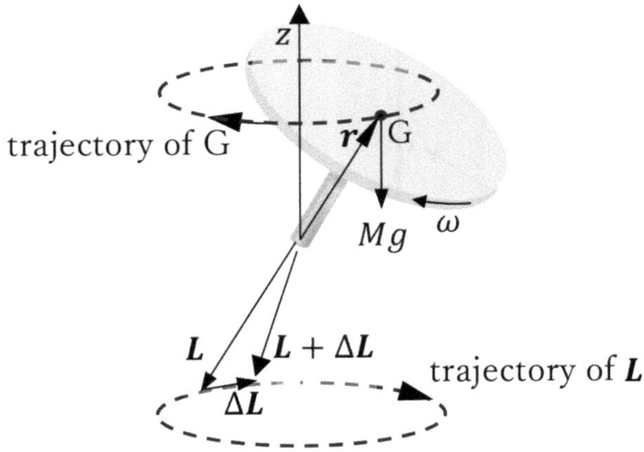

Fig. 3.12 Precession of spinning top

Now, if we write the magnitude of vector $\Delta \boldsymbol{L}$ as ΔL, we obtain the following equation as an expression of a scalar quantity.

$$\Delta L = T\Delta t = rMg\sin\theta\Delta t \tag{3.8}$$

From Fig. 3.10, the angle $\Delta\varphi$ changed by the vector ΔL is given by

$$\Delta\varphi = \frac{\Delta L}{L\sin\theta} = \frac{rMg\sin\theta\Delta t}{L\sin\theta} = \frac{rMg}{L}\Delta t \tag{3.9}$$

Since this angle is the change in the angle of the precession that occurred during the time Δt, the angular velocity of the precession Ω [rad/s] is given as

$$\Omega = \lim_{\Delta t\to 0}\frac{\Delta\varphi}{\Delta t} = \frac{d\varphi}{dt} = \frac{rMg}{L}$$

Here, if the moment of inertia of the top is J, considering a stationary motion $L = J\omega$, we get the angular velocity of the precession:

$$\Omega = \frac{rMg}{J\omega} \tag{3.10}$$

Let us apply the analytical approach we derived to a helicopter and consider the mechanism how the rotor thrust distribution leads to the aircraft tilting. Consider a helicopter with a rotor that rotates CCW when viewed from above as shown in Fig. 3.13. By using the cyclic pitch control, it is possible to make a specific distribution of lift by maximizing the pitch angle in a specific direction. Now, consider reducing the lift on the right side and increasing the lift on the left side without changing the lift on the front and rear parts of the aircraft.

Let's simplify this problem slightly and assume that an additional lift vector $\boldsymbol{F}$ [N] is generated only on the left side of the aircraft. Note that lift is assumed to be

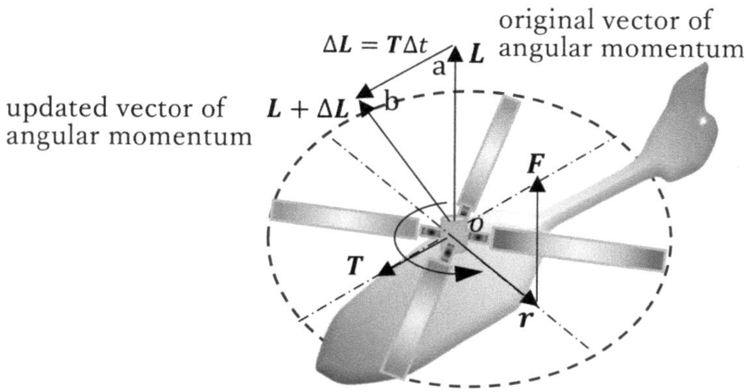

Fig. 3.13 Influence of lift control on angular momentum of rotor

generated uniformly along the length of the blade, and accordingly the vector $\boldsymbol{F}$ is drawn on the diagram at the center of the blade. The angular momentum vector $\boldsymbol{L}$ of the rotor is depicted in the direction of the rotor axis in the diagram. Instead of a gravity $-Mg\boldsymbol{e}_z$ acting as an external force on the center of gravity G in the spinning top model, the lift vector $\boldsymbol{F}$ acts as an external force in this case.

When the rotor blade has angular momentum $\boldsymbol{L}$, the external force $\boldsymbol{F}$ creates the torque $\boldsymbol{T} = \boldsymbol{r} \times \boldsymbol{F}$. Then, the change in the angular momentum $\boldsymbol{L}$ that occurs during time Δt is expressed as

$$\Delta \boldsymbol{L} = \boldsymbol{T}\Delta t \tag{3.11}$$

This vector $\Delta\boldsymbol{L}$ is drawn from the point a to point b in the diagram, and then the new angular momentum vector after Δt[s] is $\boldsymbol{L} + \Delta\boldsymbol{L}$, which is a vector from the origin O to the point b. Thus, when the maximum lift is generated on the left side of the aircraft, the rotation axis of the rotor tilts forward, and it turns out that the thrust generated by the rotor is allocated not only to lift but also to propulsive force. In other words, when a lift difference ΔF (upward force) is generated at a specific angle, an upward motion appears at a position delayed by $\pi/2$ [rad] in the direction of rotor rotation as shown in Fig. 3.14. Similarly, the reduction in lift on the right side of the aircraft $-\Delta F$ (downward force), which was omitted in the previous discussion, causes the aircraft motion at a position delayed by $\pi/2$ [rad], resulting in the aircraft to lean forward as shown in Fig. 3.14.

In addition, in the case of the top shown in Fig. 3.10, a torque acts in a direction delayed by $\pi/2$ [rad] from the position of applied gravity. It should be noted that the dynamics with angular momentum is completely different from the dynamics of a stationary object.

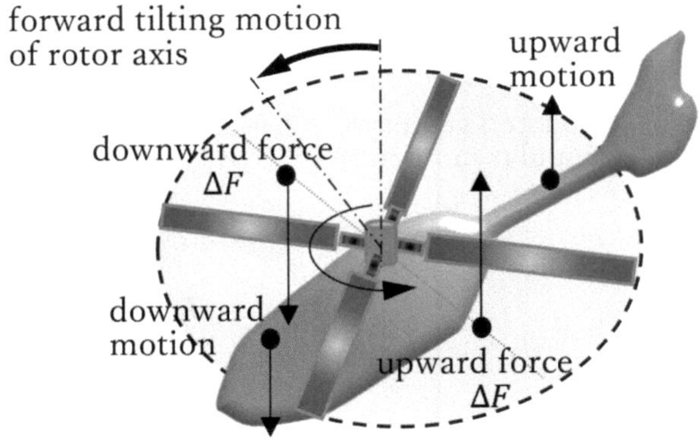

Fig. 3.14 Dynamics of cyclic pitch control. (**a**) Flight speed and blade speed. (**b**) Relative wind speed

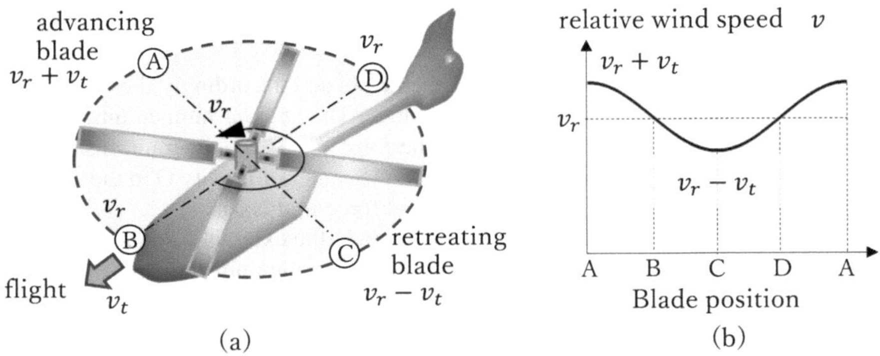

Fig. 3.15 Dynamics of cyclic pitch control

3.4.4 Relative Wind Speed Distribution of Rotor Blade

As mentioned above, the lift generated by a helicopter blade depends on the airspeed, but airspeed is determined not only by the rotation of the blade, but also by the flying speed of the aircraft. In other words, since the blade is subjected to two types of air flow. If we define the circumferential speed of the blade v_r due to rotation, which depends on the distance from the axis of rotation, and the flying speed of the aircraft as v_t, then the combined speed of the two becomes the actual airspeed v. If we consider the rotor area to be divided into the left and right areas in the direction of flight as in Fig. 3.15a, there are two areas in terms of airspeed: the area where the circumferential speed and cruising speed are added, called the advancing blade side, and the area where they are subtracted, called the retreating blade side.

At the point A in the figure, the relative wind speed in the width direction of the blade is $v_r + v_t$, and at point C, $v_r - v_t$. At the points B and D, the relative wind speed in the width direction of the blade is equal to v_r, and thus the relative wind speed changes sinusoidally depending on the position of the blade as shown in

Figure (b). In other words, the magnitude of lift, which is proportional to the square of the airspeed, becomes uneven between the left and right sides of the aircraft. Accordingly the higher the cruising speed v_t, the greater the imbalance in lift, and as a result the rotor axis is subjected to a force that changes its posture backwards.

Furthermore, if the cruising speed v_t becomes too high, the airspeed reaches the speed of sound at the position A in Figure (b), causing a significant increase in drag due to a shock wave. Conversely, the airspeed approaches 0 at the position C to get no lift force. As a result, there is a limit to the flying speed in the case of helicopter, and it is said that the theoretical flying speed limit is approximately 400 km/h, and practically approximately 300 km/h.Therefore, as flying speed increases the rotor axis is subject to the force that tilts it backward, but this uneven distribution of lift can be compensated by the flapping as described below.

In the case of tilting the rotor axis to move the helicopter forward, back, or laterally, the cyclic pitch control is employed to actively create uneven distribution of lift by means of feathering described above.

The blades rotating around the rotor axis generate lift determined by the pitch angle due to feathering, and the rotational speed. Then, flapping motion occurs influenced with gravity and air resistance (Fig. 3.16). On the other hand, we mentioned that an airspeed imbalance can cause the aircraft's posture to change, but below we describe that the air resistance affects the flapping motion to get posture stabilization.

First, the lift of the blade is proportional to the angle of attack α within a certain range of α as shown in Chap. 2, the torque τ_L [Nm] around the rotor axis created by the rotor blades is given by the following equation.

$$\tau_L \cong k_L \alpha v^2 \text{ [Nm]} \tag{3.12}$$

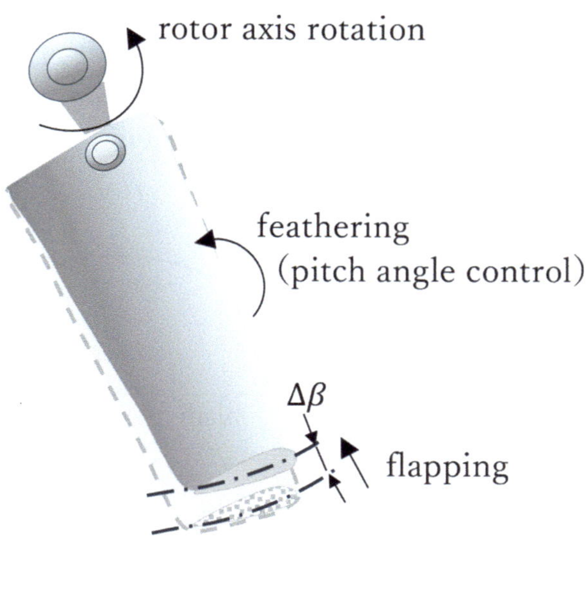

Fig. 3.16 Flapping angle change of blade $\Delta\beta$. (**a**) Torque fluctuation (when α = const.). (**b**) Flapping angle and blade posture. (**c**) Angle of attack

where k_L is a constant. We now define the load torque caused by gravity around the rotor axis as τ_g [Nm], and the damping coefficient of the torque caused by the air as ζ_b. Then the damping torque can be expressed as $\zeta_b d\beta/dt$. If we write the difference between the torques τ_L and τ_g as $\Delta\tau_L$ [Nm], then $\Delta\tau_L$ as a function of the angle of attack α and the relative wind speed v is given as

$$\Delta\tau_L(\alpha, v) = \tau_L - \tau_g = \zeta_b \frac{d\beta}{dt} \tag{3.13}$$

Now, the distribution of the torque $\Delta\tau_L$ created by the lift fluctuation before flapping occurs is determined by the set pitch angle γ, generates lift depending on the airspeed shown in Fig. 3.15, and changes for the positions A, B, C, and D (from $\theta = 0$ to $\theta = 2\pi$) as shown in Fig. 3.17a. It should then be noted here that the graph of $\Delta\tau_L$ shows the values before the blade starts flapping.

If the torque due to lift acting on the blade changes, the blade generates a flapping motion according to Eq. (3.13). As the blade performs back and forth rotational

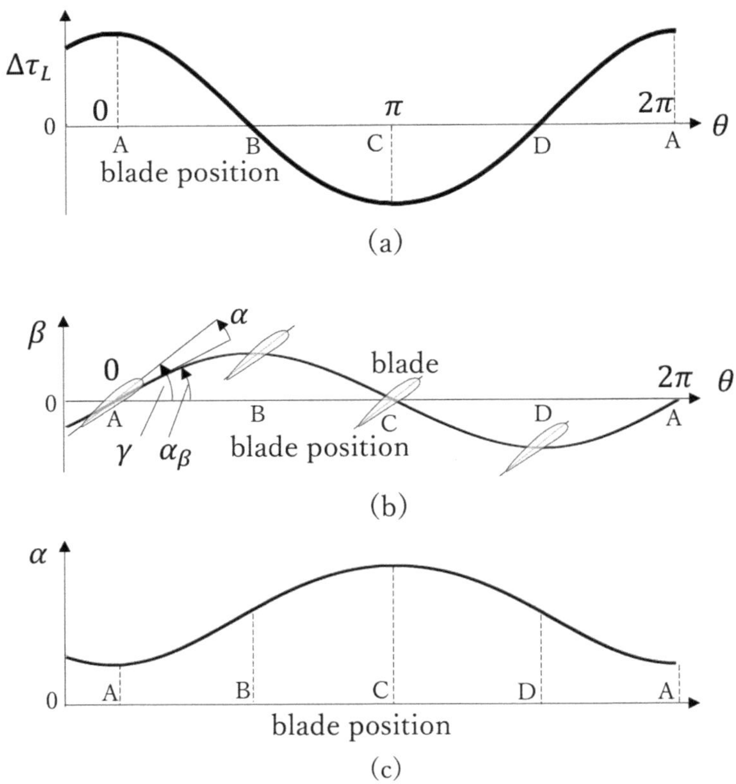

Fig. 3.17 Flapping angle change of blade

motion around the rotor axis, the direction in which it penetrates the air changes according to the flapping angle β, and the change in the angle of attack α results in a uniform distribution of lift. This is the overall mechanism. In other words, the initial lift distribution varies depending on the angular position of the blade due to differences in the relative wind speed, but flapping causes the net angle of attack to be small in areas where lift is large and large in areas where lift is small. Accordingly the lift eventually settles into a uniform distribution. Let's look at the phenomenon in detail using Fig. 3.17 below.

When the flapping angle β changes according to Eq. (3.13) as shown in figure (b), the curve of the flapping angle, which fluctuates between positive and negative, means that the angle at which the blade penetrates the air changes. Now, the angle of the blade in the direction of travel α_β is expressed by the following equation.

$$\alpha_\beta(\theta) = \tan^{-1}\frac{d\beta}{d\theta} \tag{3.14}$$

Figure 3.17b shows a blade with a pitch angle γ. If the blade generates flapping with maintaining this pitch angle, the angle of attack that penetrates the air becomes virtually zero when the angle α_β and the pitch angle γ have the same value. Therefore, the actual angle of attack α is smaller than the pitch angle γ by α_β, and we obtain the following equation for the angle of attack.

$$\alpha = \gamma - \alpha_\beta \tag{3.15}$$

The behavior of the angle of attack α is shown in Fig. 3.17c. If we assume a certain pitch angle $\gamma > 0$, then after passing $\theta = 0$ the blade is in the middle of ascending, so the angle in the direction of travel is $\alpha_\beta > 0$. And thus the angle of attack α is close to 0 in Fig. 3.17c. At the position B, the pitch angle γ has not changed, but α_β is close to 0. Then the angle of attack α is larger than at the position A, and we have $\alpha \cong \gamma > 0$. The torque due to lift with no flapping is around $\Delta\tau_L \cong 0$, but as the angle of attack α increases, the lift increases. When the blade moves to the position C, α_β now approaches its negative maximum value, and the angle of attack α reaches its positive maximum value, and $\Delta\tau_L$ was initially a negative maximum value, but instead flapping causes the lift to reach its maximum value. At the position D, which is the same as the position B, flapping increases the lift slightly. In this way, lift fluctuations occur due to the effect of airspeed as shown in Fig. 3.17a, but the angle of attack α changes from place to place as shown in Fig. 3.17c. And, in the steady state the fluctuations $\Delta\tau_L$ due to the relative wind speed are suppressed by flapping as a result.

Based on the above let's summarize the control of helicopter movement. As shown in Fig. 3.18, the degree of freedom of helicopter is 6, consisting of 3 translational motions and 3 rotational motions. The roll and pitch motions are controlled by the cyclic pitch control, while the yaw motion is controlled by varying the tail rotor output. We've also found that the cyclic pitch control causes tilting of the rotor's plane of rotation, changing the direction of thrust. And then adjusting the

Fig. 3.18 Degree of freedom

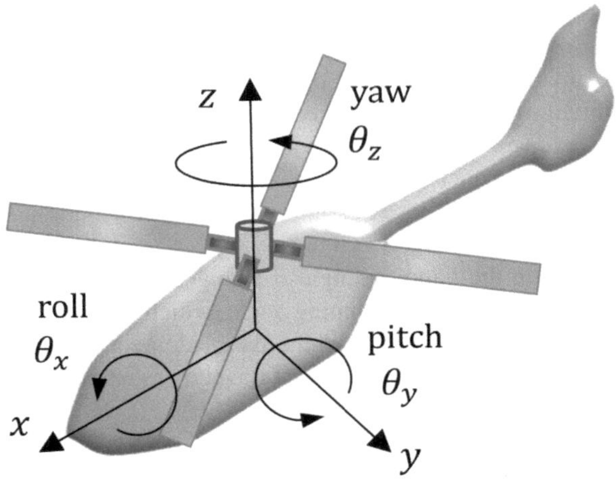

thrust with the collective pitch control varies the speed for translational movement. Furthermore, if the collective pitch control is performed when the helicopter is not tilted, the lift force can be controlled due to the changing of the main rotor thrust.

3.5 Helicopter for Exploring Mars

Figure 3.19a shows the NASA's Mars exploration helicopter "Ingenuity," which successfully flew on Mars in April 2021. The aircraft is a coaxial helicopter with two counter-rotating 2-bladed rotors, the total mass of which is only 1.8 kg. The blade area is very large, but made up of carbon fiber as shown in Figure (b). The blade shape is averaged over the entire length in terms of stress, with the chord length being shorter and the pitch angle being smaller as it moves outward. The aircraft can be said to be the size of a drone, but is a helicopter that performs collective pitch control.

Figure 3.20 shows a schematic diagram of the rotor pitch control section of Ingenuity. A pitch link for collective pitch control and a swashplate for the cyclic pitch control are attached to each lower part of the upper and lower rotors, and both rotors change lift force and control the posture. The helicopter has the rotor rotation speed of 2400 rpm, which is significantly higher than that of ordinary helicopters. The total mass is 1.8 kg that was described already, the height 0.48 m, and the total rotor span 1.2 m.

Now, we express the quantities by adding the subscript M to the characteristic quantities of Mars and E to those of Earth. The atmospheric density is $\rho_M = 0.017$ kg/m^3, the atmospheric viscosity coefficient $\mu_M = 1.13\times\ 10^{-6}$ Ns/m^2, and the gravitational acceleration $g_M = 3.73$ m/s^2. On the other hand, the atmospheric

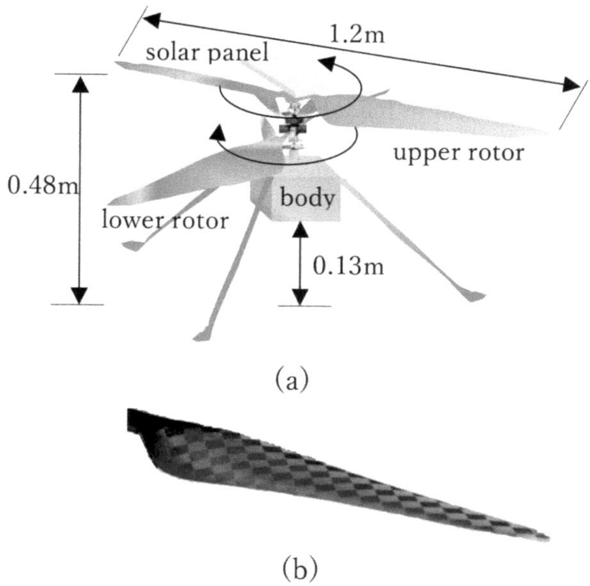

Fig. 3.19 Overview of Ingenuity. (**a**) Overview. (**b**) Carbon-fiber blade

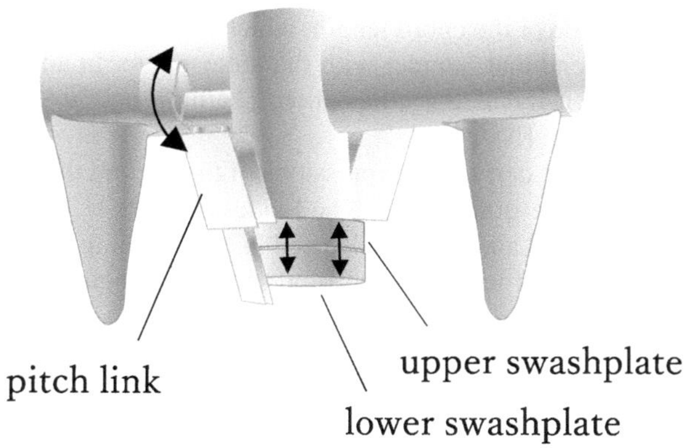

Fig. 3.20 Pitch link control structure of Ingenuity

density of Earth is $\rho_E = 1.2$ kg/m^3, the atmospheric viscosity coefficient $\mu_E = 1.79 \times 10^{-5}$ Ns/m^2, and the gravitational acceleration $g_E = 9.80$ m/s^2.

The atmosphere density on Mars is only $0.017/1.2 \cong 1/70$ of that on Earth. According to the lift equation given in Chap. 2

$$F_L = \frac{\rho_a S}{2} C_L v^2 \tag{3.16}$$

the lift is proportional to the density of the atmosphere, and thus even though the gravity is only about one-third of that, it would be significantly difficult to lift off with a conventional design.

Here, we focus on the fact that the atmosphere of Mars is characterized by its low density and extremely small viscosity coefficient. It should be noted that the atmospheric response may be different if the operating parameters on Earth are employed. In fluid dynamics, the dimensionless quantity Reynolds number is defined as the ratio between the inertial and viscous forces of the fluid. Then, the number represents the atmosphere response whether it shows laminar flow, which is a regular flow along the surface, or turbulent flow, which is an irregular flow and whether it will be in the transition region between them. Even if the size of the aircraft is different, the Reynolds number is the indicator to know the fluid flow near the surface, which will lead to ensuring the mechanical response to the blade rotation and pitch angle.

The viscous force of the atmosphere is proportional to the speed, and since it is a shear force acting through the surface in contact with the atmosphere, the viscous force is also proportional to the contact area. The inertial force in the atmosphere is the mass multiplied by the acceleration, so the Reynolds number Re is defined by dividing the inertial force by the viscous force, as given by the following equation.

$$Re = \frac{\rho V L}{\mu}$$

where L is the representative length of the object [m], such as the length of the rotor blade or the chord length, which is a length that focuses on individual physical phenomena that interact with the atmosphere. In addition, V is the airspeed [m/s], ρ the density of the atmosphere [kg/m^3], and μ the viscosity coefficient of the atmosphere [Ns/m^2].

Then we derive the expressions of the lift and the Reynolds number as a function of the blade dimensions. The blade airspeed V [m/s] is determined by the blade length L [m] and rotational speed N [rpm], being expressed as $V \propto LN$. Taking the blade's chord length as the representative length, let us compare the chord length and the airspeed on Mars and Earth, using c_M, V_M, and c_E, V_E, respectively. We obtain the following equation for the Reynolds number on Mars.

$$Re_M = \frac{\rho_M V_M L_M}{\mu_M} \tag{3.17}$$

In the case of Earth,

$$Re_E = \frac{\rho_E V_E L_E}{\mu_E} \tag{3.18}$$

As the airspeed V is proportional to the product between the blade length and the rotor speed, we have

$$\frac{V_M}{V_E} = \frac{N_M L_M}{N_E L_E} \tag{3.19}$$

And, we obtain the ratio of the Reynolds number:

$$\frac{Re_M}{Re_E} = \frac{\mu_E \rho_M V_M c_M}{\mu_M \rho_E V_E c_E} = 0.224 \frac{N_M}{N_E} \cdot \frac{L_M}{L_E} \cdot \frac{c_M}{c_E} \tag{3.20}$$

If the dimensions were the same for both cases, the Reynolds number for the aircraft traveling in Mars would be 0.224 times of the Earth. It can be seen that the rotational speed, blade length, and chord length all affect the Reynolds number proportionally. The dimensions determine the ratio of the Reynolds number in Mars compared to that of Earth.

Here the blade areas are given as

$$S_M = c_M L_M$$

$$S_E = c_E L_E$$

then we obtain the ratio of the lift:

$$\frac{F_{LM}}{F_{LE}} = \frac{\rho_M S_M}{\rho_E S_E} \left(\frac{V_M}{V_E} \right)^2 = 0.014 \left(\frac{N_M}{N_E} \right)^2 \left(\frac{L_M}{L_E} \right)^3 \frac{c_M}{c_E} \tag{3.21}$$

It is found out now that the lift is proportional to the chord length ratio c_M/c_E, proportional to the square of the rotational speed ratio $(V_M/V_E)^2$ and to the cube of the blade length ratio $(L_M/L_E)^3$.

If the mass of the aircraft for Mars is M_M [kg] and that for Earth is M_E [kg], the ratio of the required lift is

$$\frac{F_{LM}}{F_{LE}} = \frac{g_M M_M}{g_E M_E} = 0.381 \frac{M_M}{M_E} \tag{3.22}$$

Therefore, from Eqs. (3.21) and (3.22), we obtain the following equation for lift:

$$\frac{M_M}{M_E} = 0.037 \left(\frac{N_M}{N_E} \right)^2 \left(\frac{L_M}{L_E} \right)^3 \frac{c_M}{c_E} \tag{3.23}$$

This equation shows that if the aircraft for Mars had the same dimensions as the one on Earth, the mass M_M of the aircraft for Mars would be only 0.037 times that of the aircraft on Earth. This formula shows that the required mass ratio of the aircrafts depends on the rotational speed ratio, blade length ratio, and chord length ratio. Therefore the rotational speed ratio, blade length ratio, and chord length ratio can be determined based on Eqs. (3.20) and (3.23).

3.6 Drone

A drone consists of four or more rotors, mostly an even number, as shown in Fig. 3.21, and each rotor usually consists of two or three blades. A drone with four rotors is called a quadcopter, a drone with six rotors a hexacopter, and a drone with eight rotors an octocopter. Drones can take off and land vertically, or VTOL (vertical take-off and landing) just like helicopters. But their major advantages are that they are much simpler in structure than helicopters, which do not require complex parts such as swashplates, and are easier to control.

Figure 3.22 shows two typical types of quadcopters. Figure (a) is the X-type, which has X-shaped arms to the airframe, while the other shown in Figure (b) is the Plus type, which has Plus-shaped arms. In the diagram, the rotors 1 and 3 rotate counterclockwise (CCW) when viewed from above, while the rotors 2 and 4 rotate clockwise (CW), so that adjacent rotors rotate in opposite directions, canceling each other's angular momentum to prevent the body's rotation during hovering. The minimum number of rotors is four, but as the number of rotors increases, the system becomes more complicated, so that quadcopters are often used as drones. When using a heavy frame, it becomes necessary to increase the number of rotors.

Fig. 3.21 Drone

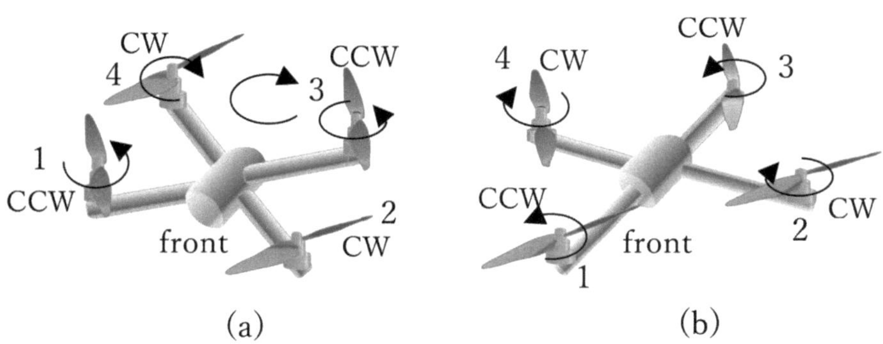

Fig. 3.22 Quadcopter. (**a**) X-type. (**b**) Plus-type

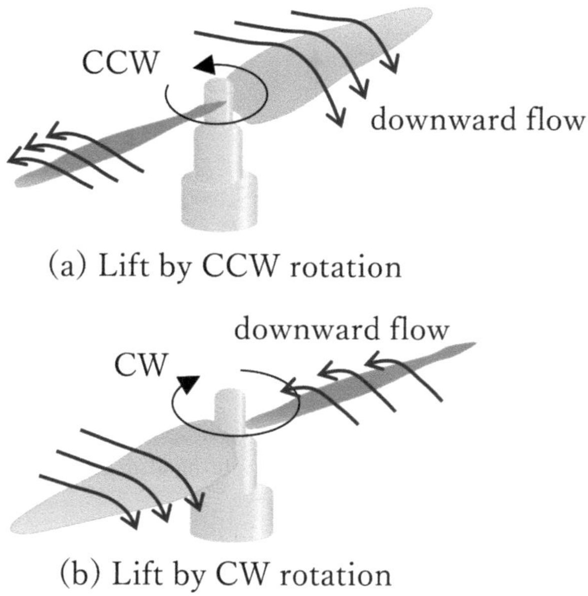

Fig. 3.23 Blades with different rotation. (**a**) Lift by CCW rotation. (**b**) Lift by CW rotation

3.6.1 Generation of Lift

Although the four rotors in Fig. 3.22 do not rotate in the same direction, they generate lift in the same direction. Using Fig. 3.23, we explain again that thrust in the same direction can be produced even with different rotation directions. In Figure (a), when the rotor rotates counterclockwise, lift is generated because the air that passes the blades is deflected downwards. On the other hand, in Figure (b), the rotor rotates clockwise, and the blades certainly create a downward airflow, generating lift in this case as well. In other words, lift can be generated even when the rotation direction is reversed by changing the shape of the blades.

Next, we consider angular momentum during hovering, and discuss how the angular momentum of a drone consisting of two groups of rotors with different rotation directions is cancelled out using Fig. 3.24, and provide a simple mathematical formula to this problem. First, it should be noticed that it is not possible to simply add up the angular momentum of each rotor: it is necessary to find the angular momentum vector with its origin at the center of gravity of the frame to take the sum of vectors.

First, if we take up one of the four rotors and consider it as a general rotating body problem, the rotational motion of the rotor with moment of inertia can be expressed as the moment of momentum $\boldsymbol{p}$ of an object whose mass point is located at a distance $\boldsymbol{r}$ from a fixed point. As shown on the right side of the figure, the

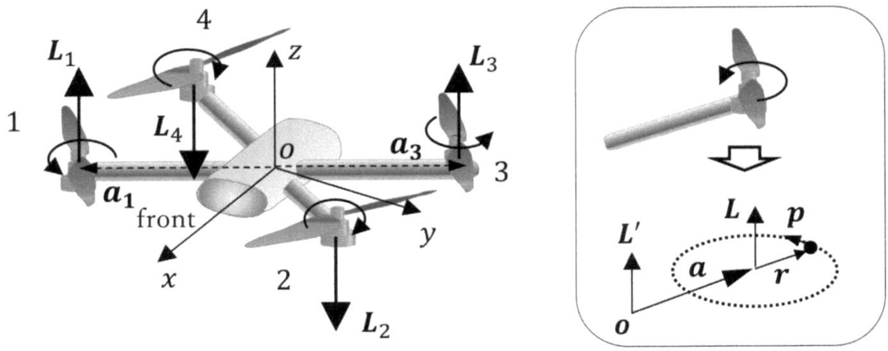

Fig. 3.24 Angular momentum produced by rotors

angular momentum $\boldsymbol{L}$ with a right-hand thread rotation about the rotation axis is given as

$$\boldsymbol{L} = \boldsymbol{r} \times \boldsymbol{p} = rp\boldsymbol{e}_s \tag{3.24}$$

where, r and p are the radius of rotation and the absolute value of the momentum, respectively, and $\boldsymbol{e}_s$ is the unit vector of the axis of rotation. In the figure, the rotors 1 and 3 rotate CCW when viewed from above, so their angular momentum vectors can be expressed as $+\boldsymbol{L}$, and the rotors 2 and 4 rotate CW, so their angular momentum vectors can be expressed as $-\boldsymbol{L}$. Now, place the origin of the (x, y, z) coordinate system at the center of gravity o of the frame, set the x-axis as the direction of travel, the y-axis perpendicular to the x-axis in the horizontal plane of the frame, and the z-axis perpendicular to the horizontal plane of the frame. If the vector from the center of gravity to the central axis of the i-th rotor is denoted as $\boldsymbol{a}_i$, then the position vector of each rotor from the center of gravity of the frame is

$$\boldsymbol{r}'_i = \boldsymbol{a}_i + \boldsymbol{r} \tag{3.25}$$

where $\boldsymbol{r}$ is the vector described in Eq. (3.24). Therefore, the angular momentum $\boldsymbol{L}_i$ of each rotor as viewed from the center of gravity coordinate o is expressed by the following equation.

$$\boldsymbol{L}'_i = (-1)^{i+1}\boldsymbol{r}'_i \times \boldsymbol{p} \quad (i = 1, \dots, 4) \tag{3.26}$$

where the sign $(-1)^{i+1}$ is positive for CCW rotation and negative for CW rotation. If the angular momentum vectors have the same origin or are on the same axis, the vectors can be summed. Therefore, when opposing rotors are located diagonally with respect to the origin, $\boldsymbol{a}_i = -\boldsymbol{a}_{i+2}$, and we have

$$\begin{aligned} \boldsymbol{L}'_1 + \boldsymbol{L}'_3 &= \boldsymbol{r}'_1 \times \boldsymbol{p} + \boldsymbol{r}'_3 \times \boldsymbol{p} = \boldsymbol{L}_1 + \boldsymbol{L}_3 = 2\boldsymbol{L} \\ \boldsymbol{L}'_2 + \boldsymbol{L}'_4 &= \boldsymbol{L}_2 + \boldsymbol{L}_4 = -2\boldsymbol{L} \end{aligned}$$

It can be seen that the sum of the angular momentum of the rotors at point symmetrical positions is equal to the sum on the same axis. Therefore, the sum of the angular momentum $\boldsymbol{L}_i'$ of the rotors with the center of gravity of the frame as the origin is

$$\sum_{i=1}^{4} \boldsymbol{L}_i' = \sum_{i=1}^{4} \boldsymbol{L}_i = 0 \tag{3.27}$$

It is found that when adjacent rotors rotate in different directions and are arranged symmetrically about a point, the sum of their angular momentum is zero. By assembling the rotors in two groups with different rotation directions and arranging them diagonally across the center of gravity of the frame, it is possible to prevent the aircraft from rotating when hovering. This is similar to a twin-rotor helicopter.

In addition, in the hovering state, $\omega_1 = \omega_2 = \omega_3 = \omega_4 \neq 0$, so if the mass of the drone is M, the gravity acceleration g, and the thrust of rotors 1 to 4 $F_i (i = 1, \ldots, 4)$, the following equation holds.

$$\sum_{i=1}^{4} F_i = Mg \tag{3.28}$$

3.6.2 Pitching and Propulsive Force

Let us consider how to generate propulsive forces. In Fig. 3.25, let us assume that the rotational speeds of the rotors 3 and 4 at the rear of the frame are $\omega_h = \omega_3 = \omega_4$,

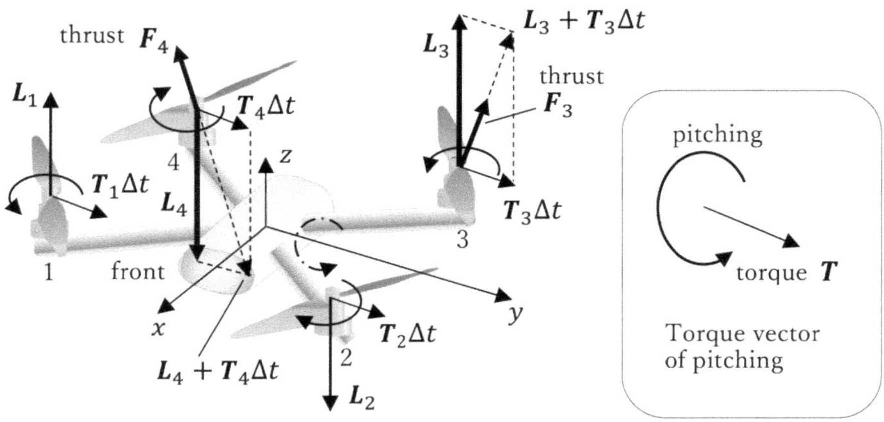

Fig. 3.25 Angular momentum and thrust in the case of pitching

and that the rotational speeds of the rotors 1 and 2 at the front are $\omega_l = \omega_1 = \omega_2$, and that temporarily $\omega_h > \omega_l$. This temporarily increases the lift at the rear, causing the frame to pitch forward, i.e., a torque is generated in the direction of a right-hand thread on the y-axis. This torque can be expressed as a vector in the positive direction of the y-axis, $\boldsymbol{T}_i (i = 1, \ldots, 4)$. As we have already learned, torque on an object due to an external force changes the angular momentum of the object itself.

The arrows of vector in the diagram are drawn of different lengths to show the magnitude difference of the angular momentum vectors. In addition, the rotors 1 and 3 rotate CCW when viewed from above, so their angular momentum vector is in the positive direction on the z axis, while the rotors 2 and 4 rotate clockwise, so their angular momentum vector is in the negative direction on the z axis.

When pitching occurs, a pitching torque $\boldsymbol{T}_i$ $(i = 1, \ldots, 4)$ acts on each rotor, which results in a change in the angular momentum vector of each rotor after Δt seconds, $\Delta \boldsymbol{L}_i = \boldsymbol{T}_i \Delta t$ $(i = 1, \ldots, 4)$. For example, if we look at the rotors 3 and 4, the vector $\boldsymbol{T}_3 \Delta t$ is added to the angular momentum vector, causing the angular momentum to change to $\boldsymbol{L}_3 + \boldsymbol{T}_3 \Delta t$ and the axis of rotation to move in the direction of that vector, resulting in the thrust of the rotor 3 being represented by the vector $\boldsymbol{F}_3 = (0, F_{3y}, F_{3z})$ in the diagram. On the other hand, the angular momentum vector $\boldsymbol{L}_4 + \boldsymbol{T}_4 \Delta t$ is shown in the diagram. Although the rotor 4 rotates in the opposite direction to the rotor 3, as we have already mentioned (see Fig. 3.23), it creates a thrust in the opposite direction to the angular momentum vector. Since the thrust of the rotor 4 is generated in the opposite direction to the angular momentum vector compared with the rotor 3, as shown by the thrust $\boldsymbol{F}_4 = (0, F_{4y}, F_{4z})$ in the diagram. And, the y-axis components of the thrust vector, F_{3y} and F_{4y}, are in the opposite directions, so that $F_{3y} + F_{4y} = 0$ to cancel out each other. The same is true for the rotors 1 and 2. The results show that the precession of the rotor blades created by the pitching torque cancel each other out, making it possible to change attitude.

Figure 3.26 shows the airframe posture after pitching motion. Then, we get the thrust $F_f = F_1 + F_2$ generated by the rotors 1 and 2, and the thrust $F_r = F_3 + F_4$ generated by rotors 3 and 4 (see Figure (b)). The rotor thrust is decomposed into

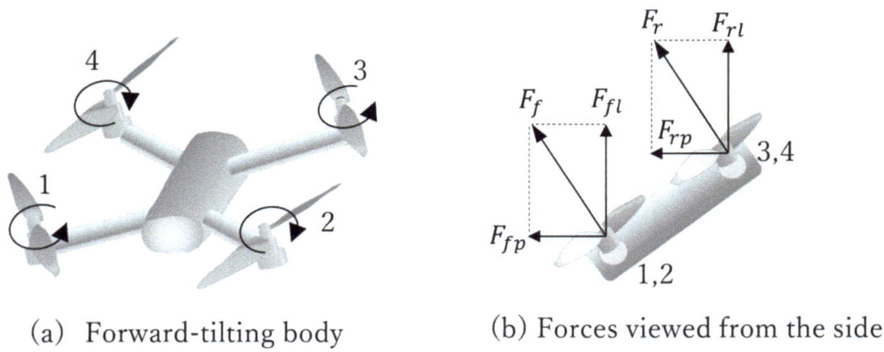

Fig. 3.26 Propulsive force. (**a**) Forward-tilting body. (**b**) Forces viewed from the side

lift and propulsive force components to get the propulsive force components F_{fp} and F_{rp}, and the lift components F_{fl} and F_{rl}. And, we obtain the total lift F_l and propulsive force F_p:

$$F_l = F_{fl} + F_{rl} \tag{3.29}$$

$$F_p = F_{fp} + F_{rp} \tag{3.30}$$

In addition, when moving backwards rather than forwards, instead of increasing the speed of the rotors 3 and 4 as described above, the speed of the rotors 1 and 2 is increased, causing the aircraft to pitch backwards and change its posture.

3.6.3 *Yawing*

In Fig. 3.27, we assume that the speed of the rotors 2 and 4 is increased, which rotate CW, where it is noted that CW rotation is equivalent to the angular momentum with a negative sign $-\boldsymbol{L}$ on the z-axis. If the diagonal pair of rotors were to have the same speed, the lift would be balanced on the diagonal, so that no tilting force would occur forward/backward or side to side. If we express the change in the angular momentum caused by increasing speed as $\delta\boldsymbol{L}$, then,

$$\boldsymbol{L}_2 = \boldsymbol{L}_4 = -\boldsymbol{L} - \delta\boldsymbol{L} \tag{3.31}$$

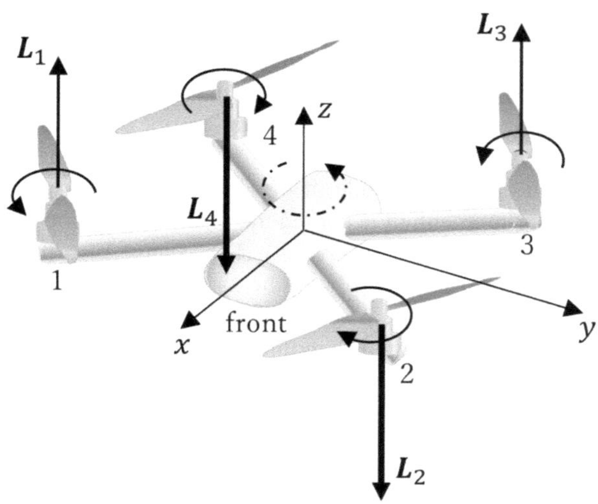

Fig. 3.27 Angular momentum for yawing

The sum of the angular momentum of the rotor as viewed at the center of gravity coordinate is equal to the one as viewed at the rotor axis, and we have the equation:

$$\sum_{i=1}^{4} \boldsymbol{L}_i' = \sum_{i=1}^{4} \boldsymbol{L}_i = -2\delta \boldsymbol{L} \tag{3.32}$$

If we now define the yawing moment of inertia of the frame part excluding the rotor as J_{yaw} [kgm^2] and the angular velocity vector as $\boldsymbol{\omega}_v$ [rad/s], the angular momentum vector is expressed as $\boldsymbol{L}_v = J_{yaw}\boldsymbol{\omega}_v$. Since the total angular momentum of the drone consisting of the frame and rotor is 0 before launch, and the angular momentum is conserved,

$$\boldsymbol{L}_v + \sum_{i=1}^{4} \boldsymbol{L}_i' = 0 \tag{3.33}$$

By increasing the speed of the rotors 2 and 4, inertia torque is generated at the airframe to get

$$J_{yaw}\frac{d\boldsymbol{\omega}_v}{dt} = 2\frac{d\delta \boldsymbol{L}}{dt} \tag{3.34}$$

causing the airframe yawing motion in the CW direction, which is opposite to the rotors 2 and 4.

3.6.4 Rolling

Let us consider the rotation about the x-axis, i.e., rolling, using Fig. 3.28. For example, the arrows representing the angular momentum vectors are drawn thicker to show the increased speed of the rotors 2 and 3. In this case, the left side of the frame is lifted by the increased lift, causing the frame to roll. And, the torque vector shown on the right side of the diagram acts on all the rotors, so that a change in angular momentum occurs in the positive direction of the x-axis, as represented by $\Delta \boldsymbol{L}_i = \boldsymbol{T}_i \Delta t$ $(i = 1, \ldots, 4)$. The change of propulsive forces generated by the rotors 2 and 3, which have the same magnitude of rotational speed, cancel each other out due to the precession, just as in the case of pitching. Similarly, the effect of the change in angular momentum is canceled out for the rotors 1 and 4. In other words, the rolling posture control can be established without being hindered by the precession.

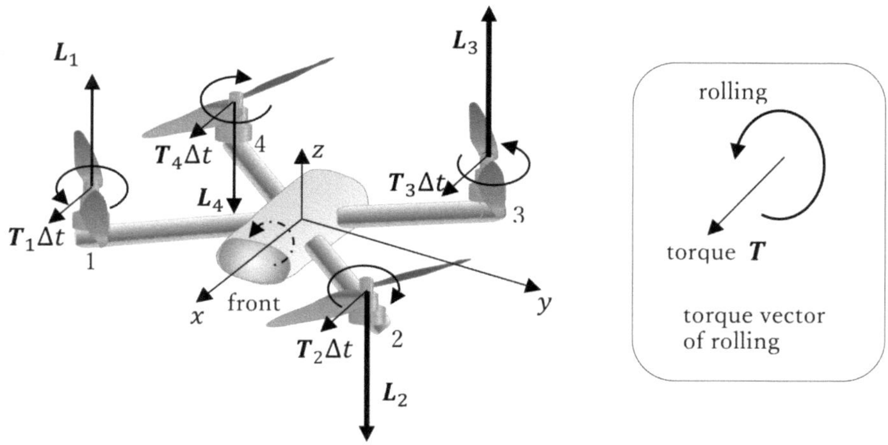

Fig. 3.28 Angular momentum for rolling

Bibliography

1. Cambridge University, Airflow across a wing. YouTube
2. D. Anderson, S. Eberhardt, How airplanes fly: a physical description of lift. The Aviation History Online Museum (2015)
3. T. Hasegawa, The plane is flying like this! (in Japanese). https://h-a-labo.amebaownd.com/posts/6415375
4. T. Benson, Incorrect Theory #1. https://www.grc.nasa.gov/WWW/k-12/VirtualAero/BottleRocket/airplane
5. D. Camuffo, Coanda effect. ScienceDirect. https://www.sciencedirect.com/topics/earth-and-planetary-sciences/coanda-effect
6. USDT, https://www.faa.gov/regulations_policies/handbooks_manualsVaviationVhelicopter_flying_handbook
7. H. Suzuki, Helicopter (in Japanese). Kodansha (2001)
8. K. Kawachi, Dynamics and control of helicopter (in Japanese). J. SICE **36**(8), 580 (1997)
9. mtk_birdman, Gyroscopic effect of helicopter (in Japanese). https://mtkbirdman.com/helicopter-aerodynamics-gyroscopic-precession
10. A. Robotics, How to create a mathematical model of a Quadcopter. https://www.youtube.com/watch?v=yIScfmQAEGY
11. NASA, MARS HELICOPTER. https://mars.nasa.gov/technology/helicopter/#Quick-Facts

Chapter 4
Levitation by Ultrasound

Abstract This chapter first describes the principle of ultrasonic transducers, then shows the derivation of the wave equation for sound waves, and reduces the solution. A quantitative explanation is given to the generation of acoustic radiation force, and then discussions are made to the acoustic levitation using standing waves and the near-field acoustic wave levitation with an object placed very close to the transducer. For the former, an analysis using the Gorkov potential is introduced to calculate the forces.

There are two types of waves, i.e., longitudinal wave and transverse wave, where each wave can be distinguished by its direction of vibration relative to its direction of travel. Longitudinal wave such as sound wave has the direction of the medium vibration in the same direction as the wave traveling, while transverse waves such as the electromagnetic waves and the waves of the liquid surface, has the vibration direction perpendicular to the traveling. Sound causes vibrations along the direction of travel including, rarefaction regions with a negative pressure fluctuation around the initial air pressure of the medium, and compression regions with positive pressure fluctuation; these are called the compression wave (Fig. 4.1a, b). Figure 4.1c shows a wave that is created when a force is applied in the longitudinal direction to one end of a spring, which is similar to the phenomenon created by the sound. But, sound cannot propagate without a medium, and thus sound cannot reach in a vacuum. It is important to note here that the compression waves propagate, though the medium only goes back and forth locally.

The characteristic variables of sound are frequency and intensity, in which the lowest frequency contained in the sound represents the pitch, and the combination of other frequency components determines the tone. Typical noise contains all frequency components from low to high frequencies at the same intensity. Human audible frequencies range from about 20 Hz to 20 kHz, and the sound above this maximum frequency is called ultrasound. The maximum acceleration α_{max} [m/s^2] of particles constituting a medium is expressed as

$$\alpha_{max} = \omega^2 A$$

T. Sakamoto, *Fundamentals of Producing Lift Forces*,
https://doi.org/10.1007/978-981-92-2222-3_4

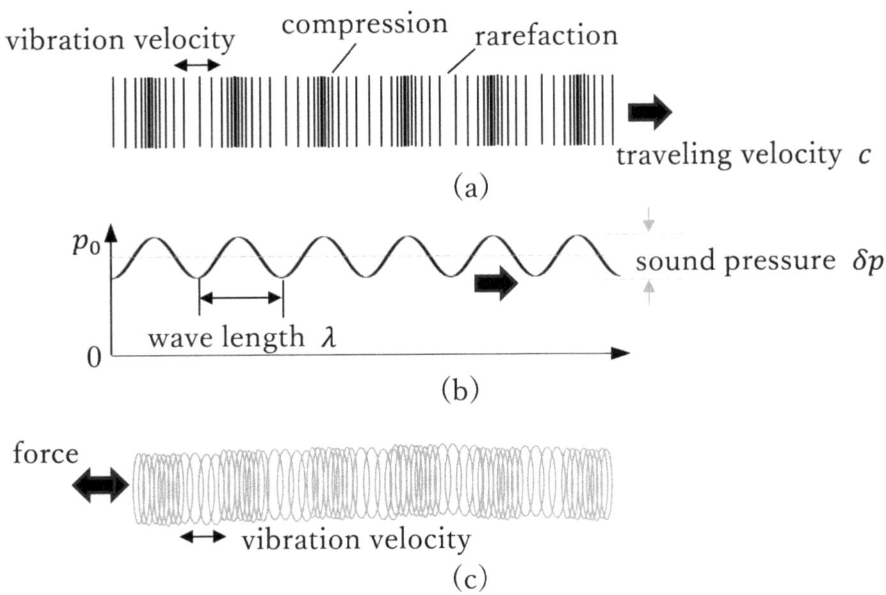

Fig. 4.1 Density and pressure of soundwave vs metal spring. (**a**) Change in density of medium. (**b**) Pressure fluctuation around the initial pressure. (**c**) Longitudinal waves generated by metal springs

where ω [rad/s] is the angular frequency ($= 2\pi f$; frequency f [Hz]), and A [m] the amplitude of the sound waves, Since the acceleration is proportional to the square of the frequency, ultrasound with high-frequency offers a significant advantage in this respect.

By utilizing the acoustic radiation force generated by ultrasound, it is possible to levitate and manipulate droplets or the force feedback in VR. The other applications include particle trapping and separation, cell trapping, plasmapheresis, and forensic analysis in biotechnology.

4.1 Ultrasound Generation by Inverse Piezoelectric Effect

When a voltage is applied to the insulator, no current flows, but microscopic charge distribution bias occurs, and a phenomenon of polarization in which charges can be seen at both ends that cannot be extracted is observed. Insulators are also called dielectrics because they exhibit such electrical phenomena. Quartz not only exhibits strong polarization but also undergoes hysteresis, and is also electrically classified as a ferroelectric material. Furthermore, when voltage is applied to quartz, significant mechanical distortion occurs. Therefore, mechanical vibrations can be generated by applying an AC voltage, and quartz was used in early ultrasound generators. Later, barium titanate (BaTiO3), which was superior to quartz, was discovered, followed

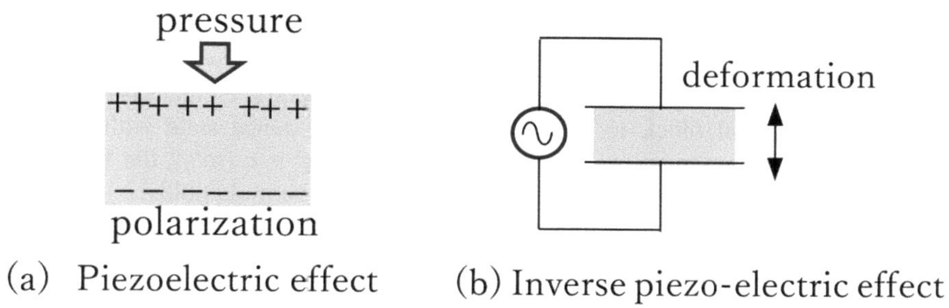

Fig. 4.2 Piezo-electric effect and inverse piezo-electric effect

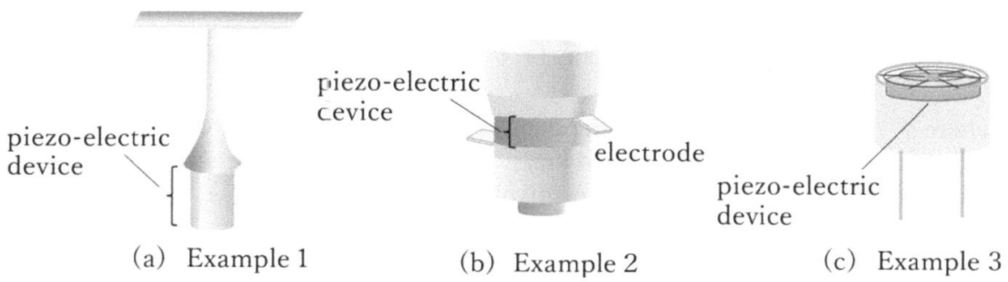

Fig. 4.3 Ultrasonic transducers and sensors

by lead zirconate titanate (PZT), leading to the emergence of ultrasonic transducers with good characteristics.

In these materials, the polarized charge appears in proportion to the strength of the applied stress, and since the charge generates an electric field, a voltage appears at both ends as a result. This is called the piezoelectric effect (Fig. 4.2a). On the other hand, when a voltage is applied to the materials, a mechanical distortion proportional to the voltage occurs as mentioned above, which is called the inverse piezoelectric effect (Fig. 4.2b). By applying an AC voltage to a piezoelectric element such as PZT, a large distortion can be obtained when the power supply frequency and the mechanical resonance frequency of the material match, which is called the piezoelectric resonance.

In addition, materials that generate a magnetic field when an external magnetic field is applied are called magnetic materials. Among them, those that have a hysteresis phenomenon are called ferromagnetic materials. Ferromagnetic materials also have a phenomenon in which mechanical distortion occurs due to an applied magnetic field as well as ferroelectric materials, which is called the magnetostriction effect. There are ultrasonic transducers that use this material, but their performance does not reach that of piezoelectric elements.

Figure 4.3 shows examples of ultrasonic transducers and sensors that use piezoelectric elements. Figure 4.3a and b shows the transducers used to emit ultrasonic waves, while Fig. 4.3c is an example of an ultrasonic sensor that can also be used to emit ultrasonic waves. The bolt-clamped Langevin-type transducer, or BLT for short, invented by the Frenchman Paul Langevin, is widely used to emit ultrasonic waves, and has a structure in which a piezoelectric element is integrated with another metal block to create a mechanical resonance state with a lower frequency. If the propagation speed of ultrasonic waves is c [m/s], the frequency f [Hz], and the wavelength λ [m], then the well-known formula holds:

$$c = f\lambda \tag{4.1}$$

In general, the propagation speed of sound through a medium depends on the type and temperature of the medium. It is fastest in solids, followed by liquids and gases. The speed in air is given by,

$$c = c_0(1 + T/273)^{1/2} \cong c_0(1 + T/546) = c_0 + 0.607T \tag{4.2}$$

where T is the temperature [°C], and $c_0 = 331.5$ m/s (the speed of sound at 0 °C). At a temperature of 20 °C, $c = 344$ m/s. The densities ρ of air and water at 20 °C are 1.20 kg/m^3 and 998 kg/m^3, respectively.

4.2 Formulation of Ultrasound Dynamics

The dynamics of ultrasound is no different from that of ordinary sound waves, and the equations developed below are not limited to ultrasound. Therefore, we will simply call them "sound waves" wherever possible.

4.2.1 Wave Equation of Sound Waves

We now consider the propagation of sound waves as a one-dimensional problem, and Fig. 4.4 shows a portion of a tube with a unit cross-sectional area. Sound waves can be expressed as the fluctuations δp [N/m^2] and $\delta\rho$ [kg/m^3] from the atmospheric pressure p_0 [N/m^2(=Pa)] and density ρ_0 [kg/m^3], respectively. And the pressure p

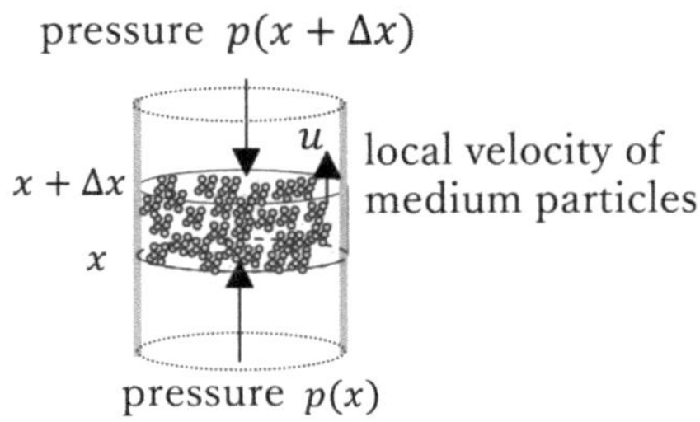

Fig. 4.4 Behaviors of medium created by sound

and density ρ are given as

$$p = p_0 + \delta p, \quad \rho = \rho_0 + \delta\rho \tag{4.3}$$

As shown in the figure, when the pressure $p(x)$ [N/m^2] acts on a medium of length Δx with a unit cross-sectional area, the local air or water particles (hereafter referred to as medium particles) move with a velocity u [m/s]. Since the mass of that part is expressed as $\rho_0 \Delta x$ [kg], the equation of motion is

$$\rho_0 \Delta x \frac{\partial u(x,t)}{\partial t} = -p(x + \Delta x) + p(x)$$

i.e.,

$$\rho_0 \frac{\partial u(x,t)}{\partial t} = -\frac{\partial p(x,t)}{\partial x} \tag{4.4}$$

Since $\partial u(x,t)/\partial t$ is the operation of differentiating the vibration velocity u with respect to t as a function of x and t, it is noted that the equation has the form $m\alpha = f$.

For a perfect fluid (or called as inviscid fluid), which has no viscosity, if we take account of an external force K acting per unit mass of the fluid, the governing equation for one-dimensional problem is generally expressed as follows:

$$\frac{\partial u}{\partial t} + u\frac{\partial u}{\partial x} = -\frac{1}{\rho_0}\frac{\partial p}{\partial x} + K \tag{4.5}$$

This is called the Euler's equation of motion. Incidentally, the operator that combines the differential operation of the left-hand side into one:

$$\frac{D}{Dt} = \frac{\partial}{\partial t} + u\frac{\partial}{\partial x}$$

is called the material derivative. The left side of Eq. (4.5) represents the acceleration of the medium particles, with the first term being the differential component of the velocity u with respect to time, and the second term showing the effect of fluid flowing into or out of the medium particles of interest due to the velocity u. Equation (4.4) can be seen as being obtained by assuming that the acceleration component caused by the local movement x of the medium particles is sufficiently small and that the system does not have an external force K.

Next, if there is an increase in mass in a medium of length Δx, this is the mass of the medium entering the region minus the mass of the medium leaving the region, so we obtain the following equation.

$$\Delta\rho\Delta x = \rho_0 \Delta t\{u(x) - u(x + \Delta x)\}$$

And, we have the following equation as a law of conservation of mass

$$\frac{\partial \rho(x,t)}{\partial t} + \rho_0 \frac{\partial u(x,t)}{\partial x} = 0 \tag{4.6}$$

This equation shows that the continuity of the flow $\rho_0 u$ is related to the change in density ρ, which represents a law of conservation of mass assuming a compressible fluid. Incidentally, Eq. (4.6) is a special case of the general form of the continuity equation with $\boldsymbol{u}$ being the velocity vector:

$$\frac{\partial \rho}{\partial t} + \mathrm{div}\rho\boldsymbol{u} = \frac{D\rho}{Dt} + \rho\mathrm{div}\boldsymbol{u} = 0 \tag{4.7}$$

and in the field of electromagnetism, it corresponds to the continuity equation that shows the relationship between the electric charge density ρ and the electric current density $\boldsymbol{j} = \rho\boldsymbol{u}$, where $\boldsymbol{j} = (j_x.j_y, j_z)$, $\boldsymbol{u} = (u_x.u_y, u_z)$, div: vector differential operator called divergence.

It is important to note here that the air is not moving as a single unit inside the tube. In other words, the density ρ simply changes over time due to local pressure changes that create a local gradient of velocity u at coordinate x along the direction of sound travel.

Using Eqs. (4.4) and (4.6), we obtain the following equation.

$$\frac{\partial^2 p(x,t)}{\partial x^2} - \frac{\partial^2 \rho(x,t)}{\partial t^2} = 0 \tag{4.8}$$

The first term on the left side represents how the slope of the pressure p with respect to the coordinate x is about to change, and the second term represents how the slope of the density ρ over time is about to change. Therefore, in the case of a monotonically linear increase or decrease, the calculation is 0, but if the slope of the change is constantly changing, such as a sine wave, it will have a non-zero value.

Figure 4.5 shows each term of Eq. (4.8) along with δp(=the variation of p), and shows that $x_1(x,t) = \partial^2 p/\partial x^2$ at coordinates (x,t) is equal to $x_2(x,t) = \partial^2 \rho/\partial t^2$ at the same coordinates (x,t). In other words, if the coordinates are the same ($x = 3L/2, t = 3T/4$) as shown in the figure, the rate of change of the gradient of p with respect to x is the same value as the rate of change of the gradient of ρ with respect to t, which matches the density and pressure of the medium in Fig. 4.1. Equation (4.8) also shows that a change in the position-wise gradient of pressure causes a change in the time gradient of density, and vice versa. From the above discussion, we can see that pressure p is a function of density ρ with time t as a parameter. Let us expand the pressure p using the infinitesimal quantity of $\delta\rho(\ll \rho_0)$ around the density ρ_0, and approximate it by

$$p(\rho) = p_0 + \delta p \cong p(\rho_0) + \left.\frac{dp}{d\rho}\right|_{\rho_0} \delta\rho \tag{4.9}$$

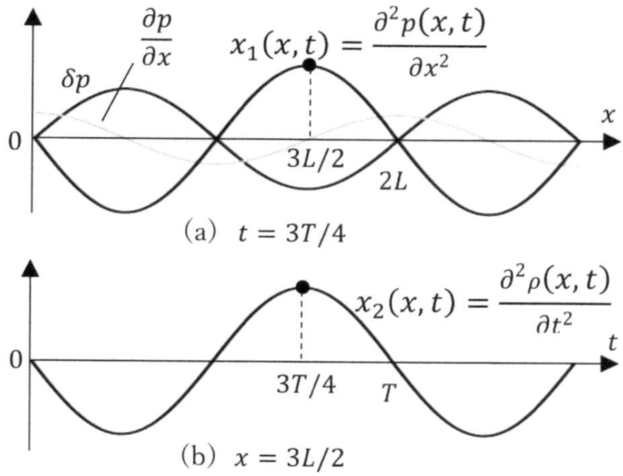

Fig. 4.5 Each term of $\partial^2 p(x,t)/\partial x^2 - \partial^2 \rho(x,t)/\partial t^2 = 0$

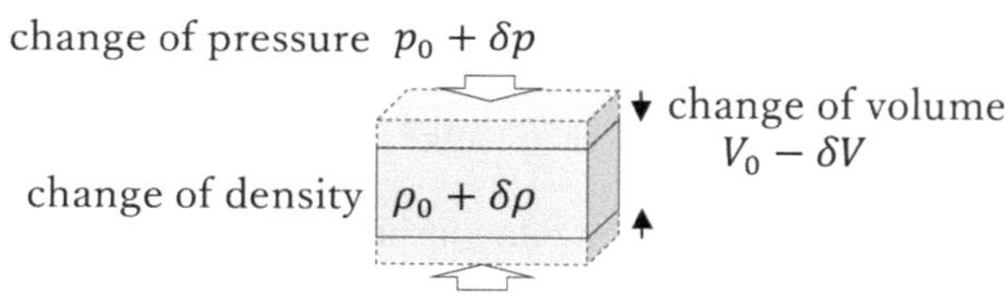

Fig. 4.6 Volume and density change by pressure

where, $p_0 = p(\rho_0)$. As shown in Fig. 4.6, when the air pressure increases from p_0 by δp, its volume changes from the initial volume V_0 by $-\delta V$, and the density increases from ρ_0 by $\delta\rho$. Since this is described as an elastic deformation, the bulk modulus of air κ [N/m^2] is defined by the following equation.In addition, fluids such as air and water do not have lateral elasticity that tends to maintain their shape, but they have volume elasticity that maintains the volume, so that only longitudinal waves occur.

$$\kappa = \frac{\delta p}{\delta V / V_0} = \frac{\delta p}{\delta\rho/\rho_0} \tag{4.10}$$

The elastic modulus of air at 20 °C is $\kappa = 1.42 \times 10^5$ [N/m^2], and that of water is $\kappa = 2.19 \times 10^9$ [N/m^2]. The reason why Eq. (4.9) is approximated to the first order of infinitesimal terms is as follows. For example, if we assume the sound of audible frequencies, the pressure fluctuation is at most about 2 Pa. Since the initial atmospheric pressure is $p_0 = 1.013 \times 10^5$ Pa, the fluctuation is only a small pressure fluctuation of less than five orders of magnitude compared to atmospheric pressure. Therefore, even a first order approximation has a sufficient accuracy. We now rewrite $dp/d\rho|_{\rho_0}$ as $\delta p/\delta\rho$ with the use of Eq. (4.10) and the constant c^2, then we have

$$\left.\frac{dp}{d\rho}\right|_{\rho_0} = \frac{\delta p}{\delta\rho} = \frac{\kappa}{\rho_0} = c^2 \tag{4.11}$$

where c is the sound wave propagation speed [m/s] given by

$$c = \sqrt{\frac{\kappa}{\rho_0}}$$

And Eq. (4.9) is expressed as

$$p(\rho) \cong p_0 + c^2 \delta\rho \tag{4.12}$$

Substituting this equation into Eq. (4.8) gives

$$\frac{\partial^2 \rho(x,t)}{\partial t^2} = \frac{\partial^2 \delta\rho(x,t)}{\partial t^2} = \frac{\partial^2 p(x,t)}{\partial x^2} \cong c^2 \frac{\partial^2 \delta\rho(x,t)}{\partial x^2}$$

And, we obtain the governing equation for the density fluctuation $\delta\rho$ [kg/m^3] related to the sound wave:

$$\frac{\partial^2 \delta\rho(x,t)}{\partial x^2} - \frac{1}{c^2}\frac{\partial^2 \delta\rho(x,t)}{\partial t^2} = 0 \tag{4.13}$$

Similarly, using the relationship $\delta\rho \cong \{p(\rho) - p_0\}/c^2$, the governing equation for the sound pressure δp [N/m^2] is given by

$$\frac{\partial^2 \delta p(x,t)}{\partial x^2} - \frac{1}{c^2}\frac{\partial^2 \delta p(x,t)}{\partial t^2} = 0 \tag{4.14}$$

Furthermore, the governing equation for the vibration velocity of the particles in the medium u [m/s] has exactly the same form:

$$\frac{\partial^2 u(x,t)}{\partial x^2} - \frac{1}{c^2}\frac{\partial^2 u(x,t)}{\partial t^2} = 0 \tag{4.15}$$

It is found out that these three physical variables $\delta\rho$, δp, and u have the same dynamics as a function of the coordinate x and the time t. The pressure fluctuation δ p in the first-order approximation case expressed by Eq. (4.3) can be calculated as acoustic pressure by specifying boundary conditions and solving Eq. (4.24).

In a three-dimensional problem, the waves propagate radially to become spherical waves. In that case, $\delta\rho$ and δp are expressed by the same type of wave equation and therefore have the same form of solution. However, the governing equation for the particle vibration velocity u of the medium is of a different form than those variables, so that the solution also has a different functional form.

4.2.2 *Solution of the Governing Equations for Density Fluctuations*

Rearranging Eq. (4.13), we have

$$\frac{\partial^2 \delta\rho(x,t)}{\partial x^2} - \frac{\partial^2 \delta\rho(x,t)}{\partial (ct)^2} = 0$$

which shows that c has the dimension of velocity. Now, if we assume that the sound source has a single angular frequency ω [rad/s], the steady-state solution is expressed as follows using the complex constant $\delta\rho^*$

$$\delta\rho = \mathrm{Re}\{\delta\rho^*(x)e^{j\omega t}\} \tag{4.16}$$

Substituting this into Eq. (4.13), we obtain

$$\frac{d^2\delta\rho^*}{dx^2} = -\frac{\omega^2}{c^2}\delta\rho^*$$

for which we get the solution using the complex constants $a_i^*(i = 1, 2)$

$$\delta\rho^*(x) = \sum_i a_i^* e^{\pm j\frac{\omega}{c}x}$$

Using Eq. (4.16), the steady-state solution is given as

$$\begin{aligned}\delta\rho =& \mathrm{Re}\left\{\sum_i a_i^* e^{\pm j\frac{\omega}{c}x} e^{j\omega t}\right\} \\ =& \delta\rho_f \cos(\omega t - kx) + \delta\rho_b \cos(\omega t + kx)\end{aligned} \tag{4.17}$$

where $\delta\rho_f$ and $\delta\rho_b$ are the constants, and $k = \omega/c = 2\pi/\lambda$ is called the wave number.

The first term on the right-hand side of Eq. (4.17) represents the sound wave traveling in the positive direction of x-axis, and the second term represents the sound wave traveling in the opposite direction. Equations (4.13)–(4.15) are called the wave equation because they represent the behaviors of waves.

Now, using the steady-state general solution expressed by Eq. (4.17), we consider the boundary conditions in the two cases of when there is a reflector to the ultrasound and when there is no reflector.

(i) Case Without Reflector If there is no reflector, then only progressive waves occur, and thus we give the following equation as the boundary condition.

$$\delta\rho|_{x=0} = \delta\rho_{fm} \cos \omega t$$

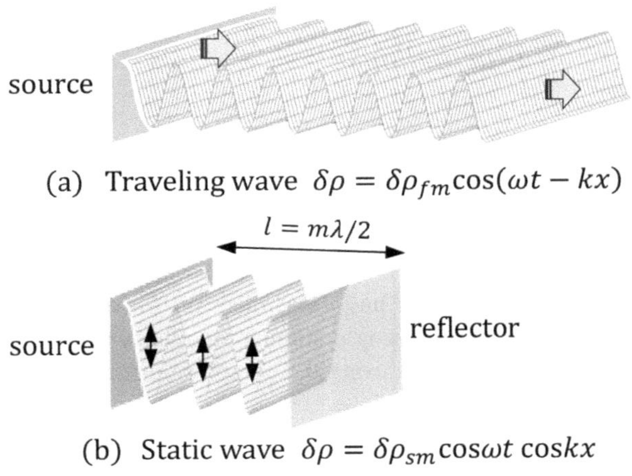

Fig. 4.7 Density change by ultrasound

where $\delta\rho_{fm}$ is a constant. Thus we have

$$\delta\rho = \delta\rho_{fm}\cos(\omega t - kx) \tag{4.18}$$

In this solution, the coordinate x with a certain constant value $\delta\rho_1 = \delta\rho|_{\omega t - kx = \theta_1}$ is expressed as

$$x = (\omega t - \theta_1)/k \tag{4.19}$$

which moves in the positive direction as time t passes. By differentiating this equation with respect to time t, we confirm that it is a traveling wave whose wave propagation speed (also called the phase velocity) is given as $dx/dt = \omega/k = c$ [m/s].

As shown in Fig. 4.7a for a plane wave (one-dimensional problem), the density fluctuation $\delta\rho$ propagates at the speed c, and the pressure also changes in the same way. From Eq. (4.12), the pressure is expressed as:

$$p(x,t) = p_0 + \delta p \cong p_0 + c^2\delta\rho_{fm}\cos(\omega t - kx) \tag{4.20}$$

It should be noticed here that the difference between the local velocity u, which depends on the coordinate x of the medium, and the wave propagation speed c. Here, the pressure fluctuation is expressed by the following equation:

$$\delta p = \delta p_{fm}\cos(\omega t - kx) \cong c^2\delta\rho_{fm}\cos(\omega t - kx) \tag{4.21}$$

where δp_{fm} represents the maximum sound pressure.

(ii) Case with Reflector When the distance l [m] between the metal plate of the ultrasound source and the reflector is an integer multiple $m\lambda/2$ $(m = 1, 2, \ldots)$ of the ultrasonic wave, the magnitudes are reinforced by reflections. In the steady state,

$$\delta\rho|_{x=0} = \delta\rho|_{x=l}$$

In Eq. (4.17), we express $\delta\rho_f = \delta\rho_b = \delta\rho_1$ by defining a constant $\delta\rho_1$, and obtain the following equation:

$$\begin{aligned}\delta\rho =&\delta\rho_1 \cos(\omega t - kx) + \delta\rho_1 \cos(\omega t + kx)\\ =&\delta\rho_{sm} \cos\omega t \cos kx\end{aligned} \tag{4.22}$$

where $\delta\rho_{sm} = 2\delta\rho_1$. This shows that the ultrasound is stationary with respect to the x-axis (Fig. 4.7b). In other words, a sinusoidal distribution $\cos kx$ is created with respect to the x-axis, and its amplitude $\delta\rho_{sm} \cos\omega t$ varies sinusoidally over time, which is called the standing wave or stationary wave. The pressure is expressed as

$$p(x, t) = p_0 + \delta p \cong p_0 + c^2\delta\rho \tag{4.23}$$

Assuming an ultrasonic wave with the frequency $f = 20\,\text{kHz}$, the wavelength is $\lambda = c/f = 0.0172\,\text{m}$ by using the propagation speed $c = 344\,\text{m/s}$ at $20\,°\text{C}$.

4.3 Acoustic Impedance and Sound Intensity

In general, sound waves actually are attenuated along the direction of propagation, and are reflected at the medium boundaries at which the propagation speed is different. However, ultrasonic waves have the advantage of having a short wavelength and less loss. In transmission lines, or distributed constant circuits, in which circuits are distributed over long distances, as typified by long-distance power transmission lines and communication lines, voltages and currents are attenuated, and are reflected at places where the impedance is different. This is similar to the situation in sound wave propagation. In other words, sound wave propagation may be formulated in a similar way to transmission lines.

Now, let us find the relationship between the pressure δp [N/m2] created by the sound wave, the density variation $\delta\rho$ [kg/m^3], and the velocity u [m/s] of the medium particles. First, the following equation holds from Eq. (4.12) as the relationship between the sound pressure and the density.

$$\delta p = c^2\delta\rho \tag{4.24}$$

This relationship holds for spherical waves in three dimensional problem.

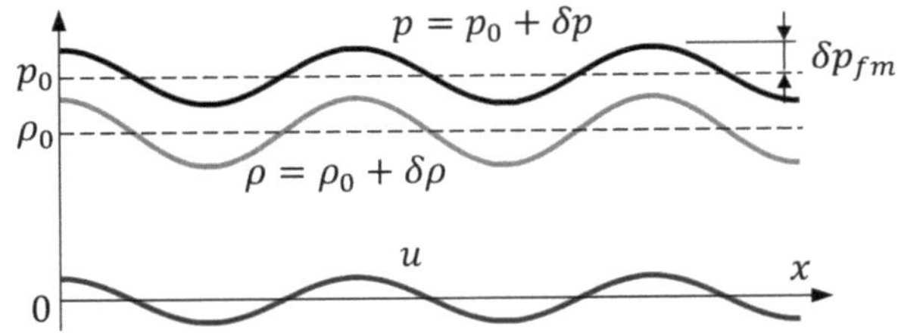

Fig. 4.8 Pressure, density and particle velocity of plane wave

Next, we consider the plane waves of a one-dimensional problem. If we substitute Eq. (4.18), which is the solution for a traveling wave, into the left-hand side of Eq. (4.6), we get

$$\frac{\partial \delta\rho}{\partial t} = -\omega\delta\rho_{fm}\sin(\omega t - kx) = -\rho_0\frac{\partial u}{\partial x}$$

then we get the relationship of the plane waves

$$u = \frac{\omega}{k\rho_0}\delta\rho_{fm}\cos(\omega t - kx) = \frac{c}{\rho_0}\delta\rho \tag{4.25}$$

where we used $c = \omega/k$. And, from Eqs. (4.24) and (4.25) we obtain the expression for the pressure δp including the medium particles velocity u:

$$\delta\rho = c^2\delta\rho = \rho_0 c u \tag{4.26}$$

That is, in the case of plane waves as a one-dimensional problem, the three variables δp [N/m^2], $\delta\rho$ [kg/m^3], and u [m/s] are expressed as functions of the same form as shown in Eq. (4.26) and Fig. 4.8. However, as already mentioned, in a three-dimensional problem, the governing equation for the velocity u of the medium particles is different from that for the sound pressure δp or density fluctuation $\delta\rho$, and it is noted that a simple relationship such as Eq. (4.26) does not hold in that case.

Now, in terms of the variables of the transmission circuit network, voltage corresponds to sound pressure δp, and current corresponds to the velocity of the medium particles u. Accordingly, the characteristic acoustic impedance z [kg/m^{2}s] is defined by the following equation.

$$z = \frac{\delta p}{u} \tag{4.27}$$

And, the acoustic impedance is defined for the cross-sectional area S [m^2]:

$$Z = \frac{S\delta p}{u}$$

Table 4.1 Properties of air and water (20 °C)

	Density ρ_0 [kg/m^3]	Elastic modulus κ [N/m^2]	Propagation speed c [m/s]	Characteristic acoustic impedance z[kg/m^{2}s]
Air	1.20	1.42×10^5	344	413
Water	998	2.19×10^9	1481	1.48×10^6

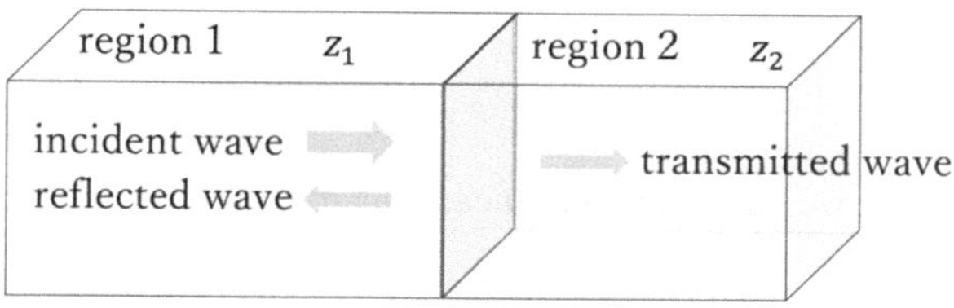

Fig. 4.9 Boundary of different acoustic impedances

At the boundary between media with different acoustic impedances, sound waves are reflected. Using Eq. (4.26), the specific acoustic impedance for a plane wave is given by:

$$z = \rho_0 c \tag{4.28}$$

Based on the formulations, the characteristics of air and water, assuming plane waves, are summarized in Table 4.1.

When two mediums with different acoustic impedance are in contact as shown in Fig. 4.9, reflection and transmission of sound waves occur at the boundary surface. The reflection coefficient R_p and the transmission coefficient T_p of amplitudes for the incident and transmitted ultrasonic waves are defined by the following equations, where z_1 and z_2 are the acoustic impedances in the region 1 on the incident side and the region 2 on the propagating side, respectively.

$$R_p = \frac{z_2 - z_1}{z_1 + z_2}, \quad T_p = 1 - R_p = \frac{2z_1}{z_1 + z_2} \tag{4.29}$$

When the characteristic acoustic impedance relation is given by $z_2 \gg z_1$, then $R_p = 1$ and $T_p = 0$, resulting in total reflection. When $z_2 = z_1$, the total transmission of ultrasound occurs expressed as $R_p = 0$ and $T_p = 1$, which is called the impedance matching.

Next, we describe the root mean square value of state variables, which is necessary to discuss the average value of sound power, etc. For example, the sound pressure and the vibration speed of the medium particles were given by

$$\begin{aligned} \delta p &= \delta p_{fm} \cos(\omega t - kx) \\ u &= u_m \cos(\omega t - kx) \end{aligned} \tag{4.30}$$

At the coordinate x of a plane wave, the sound energy of a traveling wave passing through a unit cross-sectional area in unit time, or in other words, the sound power passing through a unit cross-sectional area, is obtained by

$$i(t) = u\delta p = \frac{u_m \delta p_{fm}}{2}\{1 + \cos 2(\omega t - kx)\} \tag{4.31}$$

And, thus if we define the time average value $\langle i(t)\rangle$ of i as I [W/m^2], we obtain the following equation:

$$I = \langle i(t)\rangle = \frac{u_m \delta p_{fm}}{2} \tag{4.32}$$

Now, we define the root mean square (rms) values for the variables in Eq. (4.30) by

$$\begin{aligned} \delta P &= \sqrt{\frac{1}{2\pi}\int_0^{2\pi} \delta p^2 d\omega t} = \frac{\delta p_{fm}}{\sqrt{2}} \\ U &= \sqrt{\frac{1}{2\pi}\int_0^{2\pi} u^2 d\omega t} = \frac{u_m}{\sqrt{2}} \end{aligned} \tag{4.33}$$

Therefore, the time-averaged value of the sound power per unit cross-sectional area is given by the following equation:

$$I = U\delta P \tag{4.34}$$

which is called the sound intensity. Incidentally, the mean value of electrical power in an AC circuit, which is expressed as the product of the rms voltage and current values, needs to be multiplied by a coefficient called the power factor, which represents the phase difference between the voltage and current. However, since the sound pressure and velocity of a plane wave change with the same time phase, a phase difference coefficient is not required.

From Eq. (4.27), the specific acoustic impedance at the time t and coordinate x is given by the following equation:

$$z = \frac{\delta p}{u} = \frac{\sqrt{2}\delta P \cos(\omega t - kx)}{\sqrt{2}U \cos(\omega t - kx)} = \frac{\delta P}{U} \tag{4.35}$$

And, we obtain the sound intensity of plane waves:

$$I = U\delta P = \frac{\delta P^2}{z} = \frac{\delta P^2}{\rho_0 c} \tag{4.36}$$

By the way, since the sound power has a direction in which it propagates, the sound intensity can also be expressed as a vector quantity that has not only

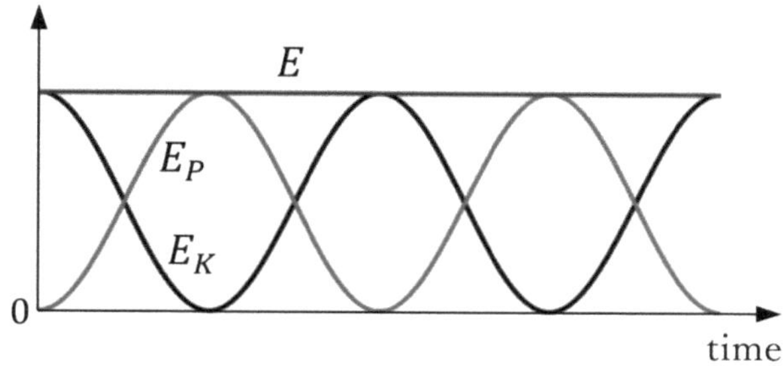

Fig. 4.10 Mechanical energies of medium particles

magnitude but also direction. Additionally, there is a method for measuring noise level by measuring the sound pressure δP, which is an intensive variable. In the mean time, representing the sound field through measuring the sound intensity I, which is an extensive variable, makes it possible to identify the sound source and separate the direct sound from reverberation.

The kinetic energy density E_K [J/m^3] per unit volume of medium particles is expressed by the following equation using the density ρ_0 and velocity u [m/s].

$$E_K = \frac{1}{2}\rho_0 u^2 \tag{4.37}$$

The total energy density E of the medium, the kinetic energy density E_K and the potential energy density E_P vary over time as shown in Fig. 4.10. When the kinetic energy density E_K is at its maximum, the potential energy density E_P is at its minimum value of 0, where the total energy density E of the medium particles is always constant. The total energy density of a plane wave is given by the equation.

$$\begin{aligned} E =& E_K + E_P = \frac{1}{2}\rho_0 u_m^2 \\ =& \rho_0 U^2 = \rho_0\left(\frac{\delta P}{z}\right)^2 = \frac{\delta P^2}{\rho_0 c^2} \end{aligned} \tag{4.38}$$

From Eq. (4.36), we get

$$I = Ec \tag{4.39}$$

which describes that the sound intensity is the flow of energy that passes through a sound propagation tube with a unit cross-sectional area in unit time.

Incidentally, sound pressure δP is measured as a relative ratio to the minimum sound pressure of hearing at the frequency of 1 kHz for a person under the best circumstances, which is $\delta P_0 = 2 \times 10^{-5}$ Pa. Since the sound pressure is an intensive variable, the squared ratio is taken to get the power ratio, and then the logarithm is taken because the response of the human follows closely to a logarithmic function:

$$L_P = 10 \log \frac{\delta P^2}{\delta P_0^2} \quad [\text{dB}] \tag{4.40}$$

The value L_P is called the pressure level. On the other hand, the ratio of the measured sound intensity I to the minimum audible value $I_0 = 10^{-12}$ [W/m^2] is expressed as

$$L_I = 10 \log \frac{I}{I_0} \quad [\text{dB}] \tag{4.41}$$

which is called the sound intensity level.

Let us consider the difference between the sound pressure level L_P and the sound intensity level L_I with the same value in terms of sound pressure. We find that the two sound levels match when the specific acoustic impedance z is 400 [kg/m^2 s]. According to Table 4.1, the characteristic acoustic impedance of air is 413 [kg/m^2 s] at room temperature, so that we confirm that the sound levels match in the air. However, we can also see that the two values are at different levels for other gases. When ultrasound is scattered by an object, acoustic radiation pressure is generated, which allows the object to levitate.

4.4 Levitation by Ultrasound

The pressure changes created by high-frequency ultrasound with a large energy density can be used for cleaning and as an acoustic speaker with super-directivity, as well as for levitating objects, although the weight is limited. Since the ultrasonic levitation does not require a container, contamination of materials can be prevented.

4.4.1 Analysis of Acoustic Radiation Pressure

In order to calculate the pressure caused by ultrasonic sound scattering, it is necessary to expand Eq. (4.9) up to second-order infinitesimal terms to get the nonlinear effect. When the sound waves are reflected at the boundary surface of an object with a different acoustic impedance from the medium, a stress component with an average value of non-zero appears. In Sect. 4.2.1, the sound pressure δp was approximated as a first-order infinitesimal quantity in pressure, but in order to calculate the effect of ultrasonic reflection, we now expand it to a second-order infinitesimal quantity to account for the nonlinear effect of the fluid. Accordingly, the pressure p, particle velocity u, and density ρ are expressed as follows.:

$$\begin{aligned} p &= p_0 + \delta p + \delta_2 p \\ u &\to u + \delta u \\ \rho &= \rho_0 + \delta\rho + \delta_2\rho \end{aligned} \tag{4.42}$$

where, the pressure p and the density ρ of the medium are expressed consisting of the first-order terms δp, $\delta\rho$ and second-order terms $\delta_2 p$, $\delta_2\rho$ in addition to the zero-order term, in which these basically becomes smaller according to the order. As mentioned above, the velocity of a medium particle does not have a zero-order term.

If we set $K = 0$, $\rho_0 = \rho$ in the Euler's equation of motion (4.5) and ignore the third-order infinitesimal quantity, we get

$$\frac{\partial \delta_2 p}{\partial x} = -\rho_0 \frac{\partial \delta u}{\partial t} - \delta\rho \frac{\partial u}{\partial t} - \rho_0 u \frac{\partial u}{\partial x} \tag{4.43}$$

Taking the time average,

$$\frac{\partial}{\partial x}\langle \delta_2 p \rangle = -\langle \delta\rho \frac{\partial u}{\partial t} \rangle - \rho_0 \langle u \frac{\partial u}{\partial x} \rangle \tag{4.44}$$

Besides,

$$\rho_0 \frac{\partial u}{\partial t} = -\frac{\partial p}{\partial x} \cong -\frac{\partial \delta p}{\partial x} \tag{4.45}$$

And, from $\delta p = c^2 \delta\rho$,

$$\begin{aligned} \frac{\partial}{\partial x}\langle \delta_2 p \rangle &= \frac{1}{\rho_0}\langle \delta\rho \frac{\partial \delta p}{\partial x} \rangle - \rho_0 \langle u \frac{\partial u}{\partial x} \rangle = \frac{1}{\rho_0 c^2}\langle \delta p \frac{\partial \delta p}{\partial x} \rangle - \rho_0 \langle u \frac{\partial u}{\partial x} \rangle \\ &= \frac{1}{2\rho_0 c^2} \frac{\partial}{\partial x}\langle \delta p^2 \rangle - \frac{\rho_0}{2} \frac{\partial}{\partial x}\langle u^2 \rangle \end{aligned} \tag{4.46}$$

Thus, using the bulk modulus of the medium $\kappa = \rho_0 c^2$, we have

$$\langle \delta_2 p \rangle = \frac{1}{2\kappa}\langle \delta p^2 \rangle - \frac{\rho_0}{2}\langle u^2 \rangle \tag{4.47}$$

The second-order pressure $\delta_2 p$ [N/m^2] generated when the ultrasound waves are scattered by an object is expressed as the sum of the first-order pressure component δp and the pressure drop effect due to the first-order velocity u of the medium particles. In this way, if there is an object that blocks the propagation of ultrasonic waves, stress in the direction of propagation acts on the object's surface. This force is called acoustic radiation force.

There are mainly two methods of achieving relatively large acoustic levitation. One is a method in which an ultrasonic transducer and a reflector are placed and a standing wave generated between them is used to levitate an object much smaller than the wavelength of the sound wave. The other method is called near-field acoustic levitation, in which the planar object is levitated by facing the ultrasonic transducer at a very close distance. The former levitates objects that are much smaller than the wavelength of ultrasound, weighing a few milligrams at most, while the latter can obtain a large levitation force by placing a planar object at a distance of several tens to hundreds of micrometers from the transducer, making it possible to levitate a planar object of several kilograms.

4.4.2 Levitation Using Standing Waves

From Eqs. (4.22) and (4.23), the standing wave sound pressure δp [N/m^2] is given by the following equation.

$$\delta p = \delta p_{sm} \cos \omega t \cos kx \tag{4.48}$$

where $\delta p_{sm} = c^2 \delta \rho_{sm}$. Standing waves can be created by placing a reflector at a position that is a multiple of half the wavelength from the sound source, or by placing a sound source in place of the reflector to further increase the sound pressure. Standing waves have antinodes, where the amplitude of the sound pressure fluctuates the most, and nodes, where the amplitude is always zero. Only very light and small objects can be levitated in the neighborhood of the node, and a restoring force acts near the equilibrium point to achieve stable levitation (Fig. 4.11). In this case, the object is limited to being much smaller than the wavelength and weighing less than a few grams. But, since the object can be levitated without using a container, this method is applied to processing that needs to avoid contamination.

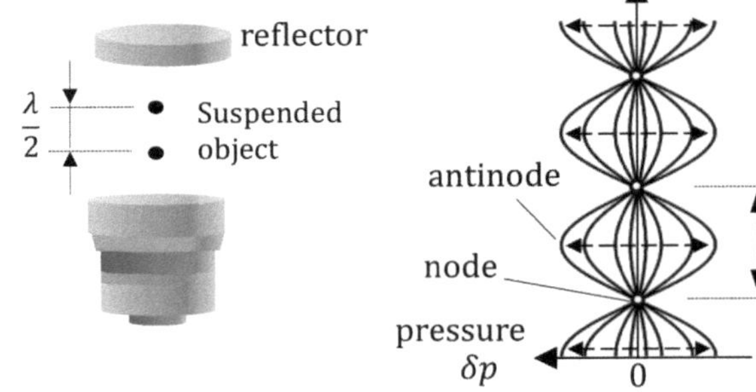

Fig. 4.11 Object trapped at node of standing wave

The Gor'kov potential method is known as a formulation of the force acting on a small sphere in a sound field. The acoustic radiation force in a standing wave is then expressed as a gradient force of the potential. When the radius R [m] of the small sphere is much smaller than the wavelength λ [m], i.e., $R \ll \lambda$, the Gor'kov potential $\Psi(x)$ [J] is defined as

$$\Psi(x) = V_s \left\{ f_1 \frac{\delta P(x)^2}{2\kappa} - f_2 \frac{3\rho_0 U(x)^2}{4} \right\} \tag{4.49}$$

where $V_s (= (4/3)\pi R^3)$ is the volume of the small sphere. Defining the density ρ_s and sound speed c_s of the sphere, the scattering coefficients f_1 and f_2 in the equations are given by

$$f_1 = 1 - \frac{\rho_0 c^2}{\rho_s c_s^2}, \qquad f_2 = \frac{2(\rho_s - \rho_0)}{2\rho_s + \rho_0}$$

For example, if the sphere is a water droplet, the density and speed of sound of water are $\rho_s = 998$ [kg/m^3] and $c_s = 1481$ [m/s], respectively. Since the density and sound speed of air are $\rho_0 = 1.20$ [kg/m^3] and $c = 344$ [m/s], respectively, the coefficients are calculated as $f_1 = f_2 = 1$. Accordingly, in that case the Gor'kov potential is expressed as

$$\Psi(x) = V_s \left\{ \frac{\delta P(x)^2}{2\kappa} - \frac{3\rho_0 U(x)^2}{4} \right\} \tag{4.50}$$

The acoustic radiation force $F_{rx}(x)$ [N] of the small sphere acting in the x-axis direction is given by the following equation:

$$F_{rx}(x) = -\frac{d\Psi(x)}{dx} \tag{4.51}$$

As the sound pressure of a standing wave is given by Eq. (4.48), using Eq. (4.4) we get

$$\rho_0 \frac{\partial u}{\partial t} = -\frac{\partial p}{\partial x} = -\frac{\partial \delta p}{\partial x} = k\delta p_{sm} \cos\omega t \sin kx$$

so that

$$\frac{\partial u}{\partial t} = \frac{1}{\rho_0} k\delta p_{sm} \cos\omega t \sin kx$$

Hence, using $k = \omega/c$ we obtain the speed of the medium particle:

$$u = \frac{1}{\rho_0 c} \delta p_{sm} \sin\omega t \sin kx \tag{4.52}$$

Here, we apply Eq. (4.33) to the standing wave to find the rms values as a function of coordinate x for the pressure fluctuation $\delta p(x,t)$ and the medium particle velocity $u(x,t)$:

$$\begin{aligned} \delta P(x) &= \sqrt{\frac{1}{2\pi}\int_0^{2\pi} \delta p(x,t)^2 d\omega t} = \delta p_{sm} \cos kx \\ U(x) &= \sqrt{\frac{1}{2\pi}\int_0^{2\pi} u(x,t)^2 d\omega t} = \frac{\delta p_{sm}}{\sqrt{2}\rho_0 c} \sin kx \end{aligned} \tag{4.53}$$

where

$$\begin{aligned} -\frac{d}{dx}\delta P(x)^2 &= \frac{1}{2}k\delta p_{sm}^2 \sin 2kx \\ -\frac{d}{dx}U(x)^2 &= -\frac{k}{2\rho_0^2 c^2}\delta p_{sm}^2 \sin 2kx \end{aligned}$$

Thus, we obtain the equation for the acoustic radiation force on a small sphere:

$$F_{rx}(x) = -\frac{d\Psi(x)}{dx} = \frac{5V_s k}{8\kappa}\delta p_{sm}^2 \sin 2kx \tag{4.54}$$

Example 4.1 (Acoustic Radiation Force and Gor'kov Potential) Consider that the radius of a water droplet with specific gravity $\rho_p = 998$ [kg/m^3] is $R = 1.5 \times 10^{-3}$ [m] to be levitated in air. A ultrasound standing wave with the frequency of 20 kHz is generated in air that has the specific gravity $\rho_p = 1.20$ [kg/m^3] and sound speed $c = 344$ [m/s]. The distance between the ultrasonic sound source and the reflector is 3/2 times the wavelength.The maximum value of sound pressure is $\delta p_{sm} = 5000$ [N/m^2].

Since the volume of the water droplet is $V_s = (4/3)\pi R^3 = 1.414 \times 10^{-8}$ [m^3], the gravity acting on the water droplet is $mg = \rho_p V_p g = 998 \times 1.414 \times 10^{-8} \times 9.80 = 1.383 \times 10^{-4}$ [N]. The angular frequency of the ultrasound is $\omega = 2\pi f = 1.257 \times 10^5$ [rad/s], the wavelength is $\lambda = 0.0172$ [m], and the wave number is $k = \omega/c = 365$ [rad/m].

Figure 4.12 uses white and black shading to illustrate the variation in density when the standing wave length created by the acoustic plane wave is $l = 3\lambda/2$ [m]. Darker areas represent increased pressure, while the center of lighter colored areas is the pressure node. The water droplet may be levitated at a position just below the node position by a distance δx [m] depending on the magnitude of its gravity.

Figure 4.13 shows the calculated results for the numerical example, where $\delta p(x,t)$ and $u(x,t)$ are shown in Figures (a) and (b) instead of $\delta P(x)$ and $U(x)$ through shading for temporal variation.

The sound pressure fluctuates with a maximum value of 5000 [N/m^2], and the maximum value of the air particle velocity is approximately 12 [m/s]. The

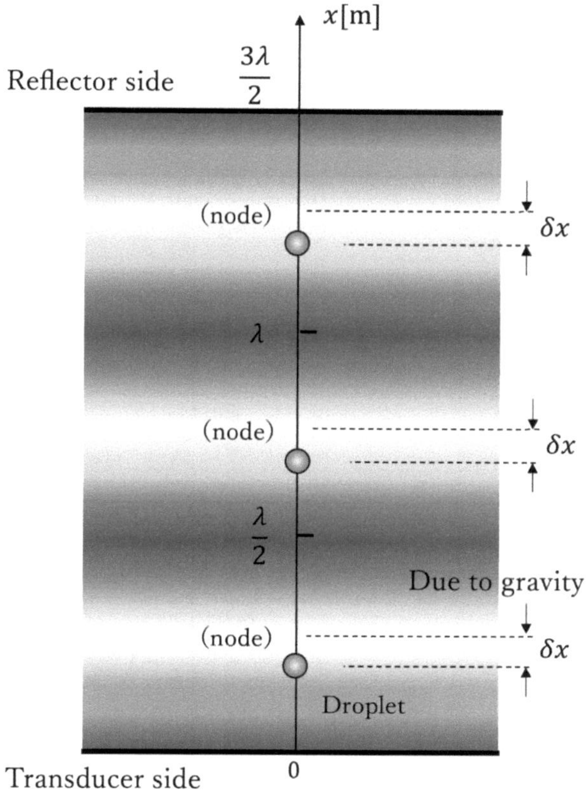

Fig. 4.12 Density caused by standing wave

Gor'kov potential has a maximum value of 6.21×10^{-7} [J] and a minimum value of -9.31×10^{-7} [J]. It can be seen that the pressure node corresponds to the minimum position of the Gor'kov potential, and its average value is negative. The first term in the parenthesis of Eq. (4.50) has a value $\delta P(x)^2/(2\kappa) \geq 0$, and the second term has a value $-3\rho_0 U(x)^2/4 \leq 0$. Accordingly, the second term, which is due to the particle velocity $U(x)$, ensures the mechanical stability of the water droplet in the x direction. In other words, since the second term lowers the average value of the potential to negative, in the beam-shaped sound field of a practical system the mechanical stability may be established even in the lateral direction because there is relatively no movement of air particles in the surrounding area. The beam-shaped sound field created by a small number of sound sources thus may provide three-dimensional mechanical stability. The generated acoustic radiation force F_{rx} has an amplitude of 5.67×10^{-4} [N], and if we write one of the stable equilibrium points of the water droplet as x_e, we get $x_e = 3.97 \times 10^{-3}$ [m] by using $F_{rx} = mg$. ◇

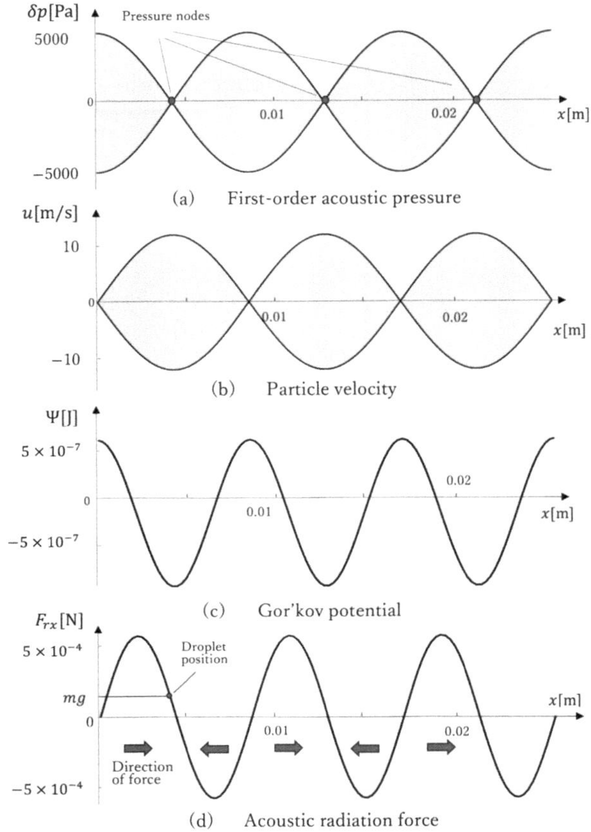

Fig. 4.13 Acoustic radiation pressure and water droplets

4.4.3 Levitation of Planar Object Using Acoustic Radiation Pressure

When there is an object that blocks the propagation of ultrasound, stress acts on the surface of that object in the direction of propagation. In other words, sound waves are reflected at the boundary surface of the object, which has a different acoustic impedance from the medium, and a stress component appears whose average value is not zero. This is called the acoustic radiation pressure, and the resulting force is called the acoustic radiation force. In the previous section, the sound pressure δp was approximated as a first-order infinitesimal quantity in pressure. However, if the approximation is further expanded to a second-order infinitesimal quantity by analyzing an equation that takes into account the nonlinear effects of fluid, the acoustic radiation pressure can be found.

When an object is placed opposite the ultrasound source as shown in Fig. 4.14, the acoustic radiation pressure P_r [Pa] expressed as a time-averaged value and the

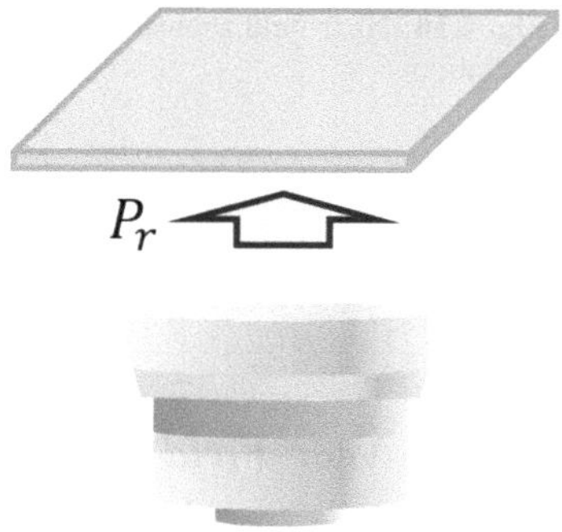

Fig. 4.14 Levitation by acoustic radiation pressure

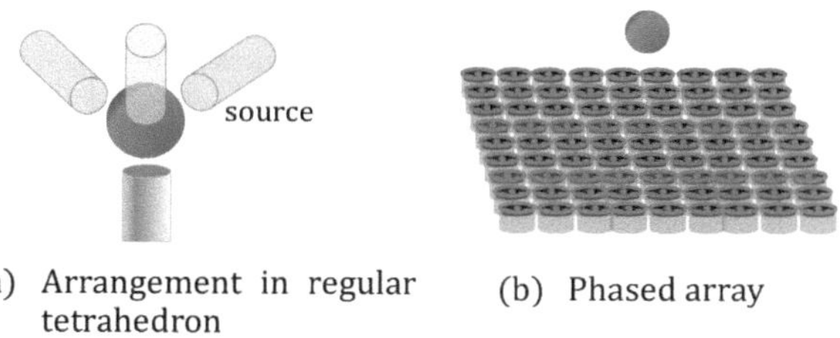

Fig. 4.15 Levitation using multiple transducers

acoustic radiation force F_r [N] acting on an object with a radiated area S [m^2] are given by the following equations:

$$\begin{aligned} P_r =& \alpha_D E = \alpha_D \frac{\delta P^2}{\rho_0 c^2} \\ F_r =& S P_r = \alpha_D S \frac{\delta P^2}{\rho_0 c^2} \end{aligned} \tag{4.55}$$

where $\alpha_D = 1 + R_p - T_p$ is the coefficient of the force caused by the reflection of ultrasound determined by the surface of the object, and E [J/m^3] is the energy density of the ultrasound. In the case of total reflection, $\alpha_D = 2$, while in the case of total transmission $\alpha_D = 0$.

Figure 4.15a shows an example of levitating a very light object using traveling waves generated by four ultrasound sources. Static stability of levitation is ensured by placing the transducers at the vertices of a regular tetrahedron, and the three-dimensional position of the object is controlled by adjusting the magnitude and duration of each ultrasonic wave output. Figure 4.15b shows a system in which several hundred or more transducers are arranged in a planar fashion. By individually controlling the phase of the sound waves, it is possible to control the propagation direction and concentrated radiation of the synthesized ultrasonic waves; this system is called the ultrasonic phased array.

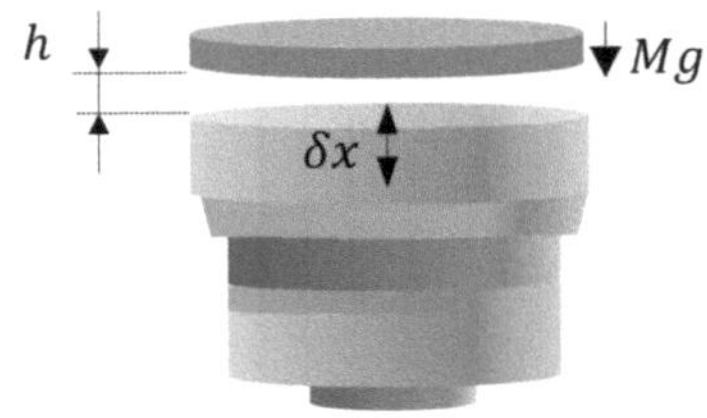

Fig. 4.16 Near-field acoustic levitation

4.4.4 Near-Field Acoustic Levitation

If a planar object is placed very close to the sound source so as to almost reflects sound waves, the object may act as a reflector, exerting a large acoustic radiation pressure to levitate the object as shown in Fig. 4.16. This is called the near-field acoustic levitation. Suppose that the surface of the transducer vibrates being given by the equation:

$$\delta x = \Delta h \cos \omega t \tag{4.56}$$

When the levitation height is h [m], the levitation force per unit area acting on the object, P_r [N/m^2], is given by the following equation:

$$P_r = \frac{1+\gamma}{4} \rho_0 c^2 \frac{\Delta h^2}{h^2} \tag{4.57}$$

where γ is the specific heat ratio of the medium, which is approximately 1.4 for air. Stable levitation requires an area with dimensions approximately equal to or larger than the wavelength. Furthermore, because near-field acoustic levitation utilizes a confined space that transmits sound waves, it is multiplied by the Rayleigh radiation pressure coefficient $(1+\gamma)/2$, which corresponds to the conditions for a medium inside a sealed container.

Example 4.2 (Levitation by Ultrasound) The amplitude of the transducer's vibration surface is expressed as $\Delta h = 5\,\mu\text{m}$, and the levitating body with the area of $S = 0.01\,\text{m}^2$ is placed at a height of $h = 50\,\mu\text{m}$. Calculate the levitating force in the air assuming that the surface of the levitating body totally reflects the ultrasound.
Solution The levitating force per unit area is

$$f_L = \frac{1+1.4}{4} \times 1.20 \times 344^2 \frac{(5\times 10^{-6})^2}{(5\times 10^{-5})^2} = 852\,\text{N/m}^2$$

And, we obtain the levitation force

$$F_L = SP_r = 0.01 \times 852 = 85.2\,\text{N}$$

The result shows that it is possible to levitate a mass of approximately 0.87 kg (= 8.52/g). Furthermore, if an object weighing 3.5 kg, which is about four times the mass, is placed instead, the levitation height becomes half that of the original height, $h = 25\,\mu\text{m}$. ◇

Bibliography

1. K. Mizukoshi, Fundamentals of ultrasound (in Japanese). http://polar.imr.tohoku.ac.jp/_userdata/sono34.pdf
2. Y. Ito, Methods for generating high-power airborne ultrasound and their application technology (in Japanese). https://www.jstage.jst.go.jp/article/essfr/9/3/9_205/_pdf/-char/ja
3. M. Sano, Acoustic radiation pressure of airborne ultrasound (in Japanese). Toin J. (38), 131 (2018)
4. A. Omoto, Understanding acoustics from the wave equation. J. Acoust. Soc. Japan **66**(9), 541 (2010)
5. A. Kozuka, H. Yasui, Study of sound field for ultrasonic levitation. http://www.acelec.cei.uec.ac.jp/nl/rec/NL2007/NL-07-09.pdf
6. S. Koyama, Propagation of sound waves (2021). https://sp.ipc.i.u-tokyo.ac.jp/~koyama/teaching/appl_acoust_2021/appl_acoust_day1.pdf
7. T.F. Hueter, R.H. Bolt, *SONICS; Techniques for the Use of Sound and Ultrasound in Engineering and Science* (Wiley, 1955)
8. Y. Hashimoto, Y. Koike, S. Ueba, Non-contact object transportation using near-field acoustic waves. J. SICE **38**(2), 105 (1999)
9. K. Nakamura, Transportation of flat objects by near-field acoustic levitation. J. Acoust. Soc. Japan **69**(11), 603 (2013)
10. H. Hatano, Ultrasonic levitation and its application to device-free processing. Jpn. J. Appl. Phys. **67**(3), 298 (1998)
11. K. Suda, A. Kaneko, K. Hasegawa, Y. Abe, Follow-up transport of levitated droplets using counter-type focused ultrasound. Jpn. J. Multiphase Flow **35**(1), 68 (2021)
12. T. Hoshi, Development of a compact acoustic focusing device that generates non-contact forces. J. SICE **50**(7), 543 (2014)
13. M.A.B. Andrade, N. Prerez, J.C. Adamowski, Review of progress in acoustic levitation. Gen. Appl. Phys. **48**, 190 (2018)
14. M. Settnes, H. Bruus, Forces acting on a small particle in an acoustical field in a viscous fluid. Phys. Rev. **E85**, 016327 (2012)

Chapter 5
Levitation by Electric Field

Abstract The chapter describes the principle of levitation using electric field, and explains that there is a maximum value of electric field that causes dielectric breakdown, which limits the generated forces. The mechanism and characteristics of dielectric breakdown are described.

When a certain type of matter is subjected to a force when placed in a certain space, that space is called the "field". Thus, a field in which a matter with mass is subjected to a force is called the gravitational field, and a field in which a matter with an electric charge is subjected to a force is called the electric field. A field in which an electric current or magnetic material is subjected to a force is called the magnetic field. We experience these three types of fields in our daily lives, all of which act on matters from a distance, and have in common that the force they exert is proportional to the inverse square of the distance (Fig. 5.1). The force created by electric and magnetic fields is collectively called electromagnetic force, and the significant difference between this and the gravitational field is the controllability of the field strength.

When creating a levitation system using an electric or magnetic field, the first issue to be addressed is the system stability. Stable levitation requires, (1) static stability and (2) dynamic stability. Figure 5.2 shows a conceptual diagram. First, a mechanical equilibrium point must exist, and the system is said to be statically stable when there is a restoring force that moves the object toward the equilibrium point (Fig. 5.2a). However, static stability alone is not sufficient. If the friction between the object and the moving surface in the figure can be ignored, the object may not stop at the equilibrium located at the lowest point. The potential energy that is given to the object at the beginning is partially converted to kinetic energy in the next instant and the object begins to move, and the next moment it reaches the highest point, the kinetic energy becomes 0, and only potential energy remains. The total energy, which is the sum of potential energy and kinetic energy, is conserved and not consumed. However, if friction occurs between the object and the surface, the energy is consumed as frictional energy over time, and the motion decays over

T. Sakamoto, *Fundamentals of Producing Lift Forces*,
https://doi.org/10.1007/978-981-92-2222-3_5

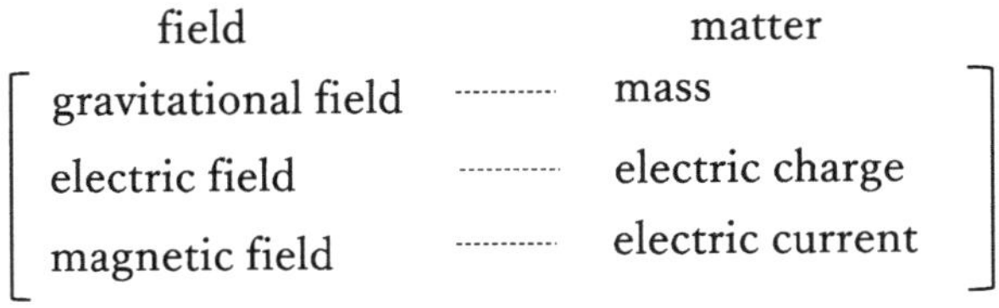

Fig. 5.1 Field vs matter to receive force

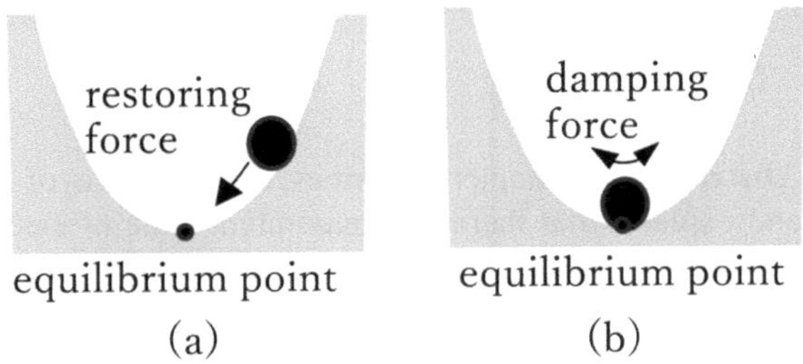

Fig. 5.2 Stability. (**a**) Static stability. (**b**) Dynamic stability

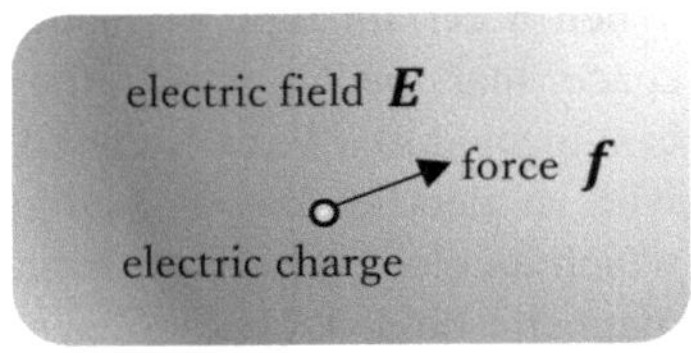

Fig. 5.3 Knowing electric field by electric charge

time. Therefore, it is not enough to ensure static stability; dynamic stability is also required (Fig. 5.2b).

Now, an electric field is defined by the force acting on an electric charge, and this force is expressed using the coordinate system (x, y, z) as $\boldsymbol{f} = (f_x, f_y, f_z)$[N]. We assume a point charge q (Fig. 5.3) to see how the force acts. Since the force is proportional to the amount of charge, the strength of the field can be calculated by dividing the force by the amount of charge, $\boldsymbol{E} = \boldsymbol{f}/q$. This field is called the electric field. A mechanical system driven by the electric field does not even have a statically stable equilibrium point according to Earnshaw's theorem, so that a control system must be constructed to obtain a stable levitation system.

5.1 Response of Conductors and Insulators to Electric Fields

Figure 5.4 shows two cases where a conductor and an insulator are sandwiched between electrodes with applied voltages. The term dielectric is used instead of insulator since we here focus on the electrical reaction of the insulator. The sample with a thickness of h [m] is separated from the electrodes by a small gap $x_0/2$ [m] above and below, and there is no direct electrical connection between the electrodes and the sample. Now, as shown in the figure, electric charges appear on the surface of both samples, but there is a significant physical difference.

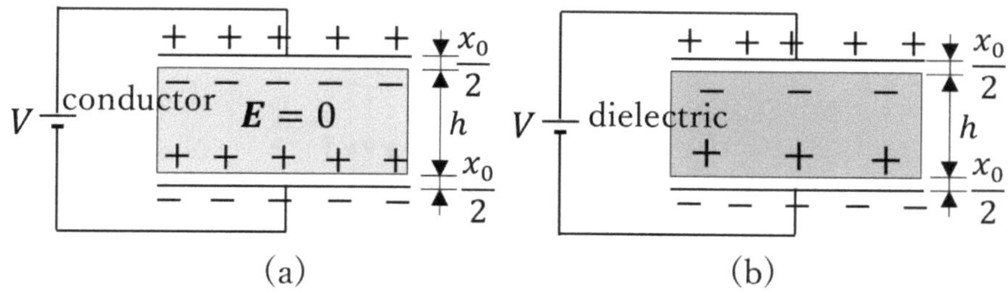

Fig. 5.4 Responses of conductor and dielectric material. (**a**) Conductor. (**b**) Dielectric material

In both cases, when a power supply voltage V is applied, free electrons move in the circuit, bringing negative charges on the cathode, and positive charges on the anode. When the potential difference created by the electric field generated between the electrodes becomes equal to the power supply voltage, the movement of free electrons, and therefore the current, becomes zero. However the situation in which the charges generated on the electrodes create in the conductor and the dielectric is different. In the case of the conductor sample shown in Fig. 5.4a, the free electrons inside move macroscopically to generate a current, as long as the electric field is not zero. When a distribution of free electrons is achieved inside the conductor such that the electric field strength $\boldsymbol{E} = 0$, a charge distribution is created on the surface of the conductor that cancels out the effect of the charge on the electrode; this is called the electrostatic induction.

On the other hand, in the case of a dielectric sample as shown in Fig. 5.4, there are no free electrons; instead, there is a distribution of positive and negative charges in the ion or molecular level. In this case, macroscopic movement of electric charge is not observed; the relative distance between positive and negative charges only shifts microscopically. When this microscopic shift in positive and negative charges occurs overall, charges appear on the surface of the dielectric. This phenomenon is called the polarization. The polarization charges that appear on the sample surface in the case of a dielectric is smaller than the charges generated by electrostatic induction in a conductor, where another major difference is that $\boldsymbol{E} \neq 0$ inside a dielectric sample.

5.2 Force Between Electrodes

We here ignore the curvature of the electric field at the ends of the electrodes and assume that the electric field is uniform in strength everywhere between the electrodes. We give a simple explanation of the concept of Maxwell's stress using Fig. 5.5, and afterwards apply it to the derivation of the force between electrodes in Fig. 5.4. Assume that an electric field $\boldsymbol{E}$ [V/m] is directed from top to bottom in the figure with a uniform strength over a cylindrical part of the area S [m^2] in a space

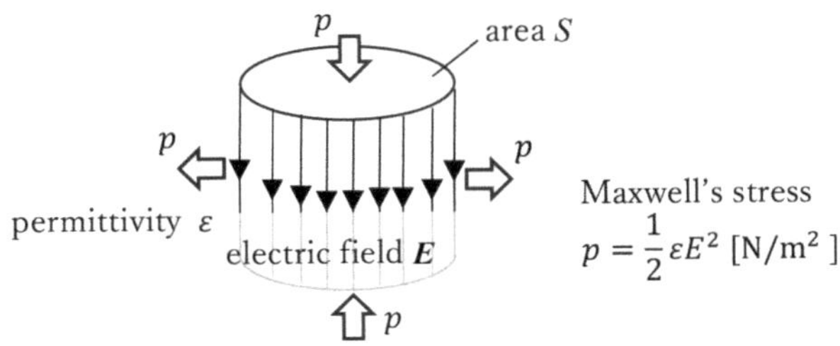

Fig. 5.5 Stress of electric tube of force

with the permittivity ε. The absolute value of $\boldsymbol{E}$ is expressed by the line spacing in the diagram, while the direction of $\boldsymbol{E}$ is represented by an arrow, which is called the line of electric force. The figure shows a tube bordered by electric field lines, which is called the electric tube of force. The electric tube of force is characterized by the properties that shows a stress $p = \varepsilon E^2/2$ [N/m^2] to cause shrinking of the tube in the longitudinal direction, and a stress $p = \varepsilon E^2/2$ [N/m^2] expanding the tube in the lateral direction. This is called the Maxwell's stress.

Now, in the steady state of Fig. 5.4a, the electric field intensity inside the conductor is 0. Thus, for the applied voltage V [V], the total length over which the electric field occurs is only x_0 [m], and we obtain the following equation.

$$V = x_0 E \tag{5.1}$$

Therefore, the attractive force F [N] between the electrode and the conductor sample in Fig. 5.4a is obtained by multiplying Maxwell's stress p by the area S:

$$F = pS = \frac{1}{2}\varepsilon_0 E^2 S = \frac{\varepsilon_0 S}{2x_0^2} V^2 \tag{5.2}$$

where ε_0 is the permittivity of air.

In the case of Fig. 5.4b, when an electric field is applied to the dielectric from the outside, the electric field decreases in inverse proportion to the relative permittivity $\varepsilon_r = \varepsilon/\varepsilon_0$. Therefore, the relationship between the supply voltage V and the electric field intensity E in the gap is

$$V = \left(x_0 + \frac{h}{\varepsilon_r}\right) E \tag{5.3}$$

Then, the electric field intensity E [V/m] in the gap and the electric field intensity E_d [V/m] inside the dielectric are given by the following equations.

$$\begin{aligned} E &= \frac{V}{x_0 + \frac{h}{\varepsilon_r}} \\ E_d &= \frac{V}{\varepsilon_r x_0 + h} \end{aligned} \tag{5.4}$$

Therefore, the attractive force acting between the electrodes and the dielectric sample is given by

$$F = \frac{1}{2}\varepsilon_0 E^2 S = \frac{\varepsilon_0 S}{2(x_0 + h/\varepsilon_r)^2} V^2 \tag{5.5}$$

We have formulated the attractive force between the electrodes and the samples for the two cases: a conductor and a dielectric as a sample. Comparing Eqs. (5.2) and (5.5), only the term relating to ε_r in the denominator is different, which makes the attractive force of the dielectric smaller. In other words, the reason for the larger attractive force is that the charge that appears in the conductor is larger than that of the dielectric. However, when the strength of the electric field changes over time, the movement of free electrons occurs in the case of a conductor, causing current flows to generate Joule losses, which leads to a major disadvantage as an actuator. On the other hand, in the case of a dielectric some dielectric loss occurs depending on the material quality.

We obtained the quantitative relationship of the attractive force created by the electric field between the electrode and the sample for the two cases. From these discussions, we realize that non-contact levitation is possible for both conductors and dielectrics, and this type of levitation is called the electrostatic levitation.

Incidentally, non-contact levitation is required in processes such as semiconductor wafers. Figure 5.6 shows an example of an electrostatic levitation system. When levitating glass or silicon, for example, if electrodes are placed with a gap x_0 as shown in the figure, an attractive force is generated between the electrode and the sample as mentioned above. However, this is an unstable system that does not satisfy both static and dynamic stability. Accordingly, stabilization is required by controlling the power supply voltage based on the information on state variables such as the gap signal.

Example 5.1 (Supply Voltage Required for Electrostatic Levitation) In Fig. 5.6, the gap $x_0/2 = 0.005\,\text{m}$, the area of the electrode $S = 0.0179\,\text{m}^2$, the mass, thickness and relative permittivity of the sample are $m = 1\,\text{kg}$, $h = 0.01\,\text{m}$ and $\varepsilon_r = 10$, respectively, and the gravitational acceleration $g = 9.80\,\text{m/s}^2$. The power supply voltage V [V] and the intensity of the electric field required for levitation are estimated.

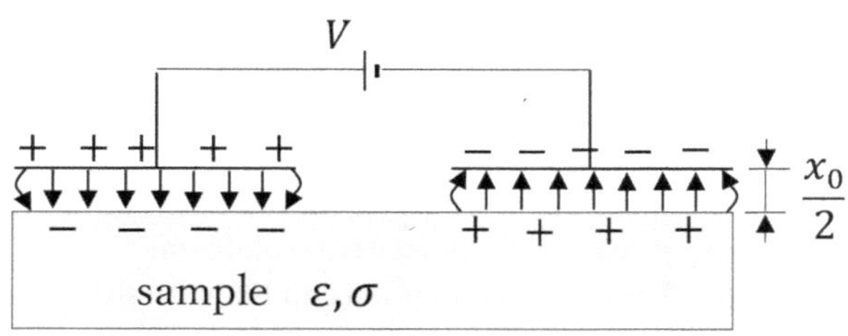

Fig. 5.6 Electrostatic levitation

Solution The relationship between the power supply voltage and the electric field intensity in the gap is approximated by the following equation:

$$V \cong x_0 E$$

And, the attractive force is given by

$$F = 2 \times \frac{1}{2}\varepsilon_0 E^2 S \cong \frac{\varepsilon_0 S}{x_0^2} V^2$$

Since the attractive force becomes equal to the gravity force in the steady state, we have

$$9.80 \cong \frac{8.85 \times 10^{-12} \times 0.0179}{0.01^2} V^2$$

Thus we obtain the required voltage of $V = 78.7\,\text{kV}$, giving the required electric field intensity of $E = 7.87 \times 10^6\,\text{V/m}$. However, this exceeds the threshold electric field intensity of $3 \times 10^6\,\text{V/m}$, which is the electric field that initiates dielectric breakdown in air. Therefore, in such a case, levitation is limited to a vacuum environment. ◇

5.3 Dielectric Breakdown

Assuming electrostatic levitation in air, we are faced with the aforementioned problem of dielectric breakdown. The dielectric breakdown in air is caused by air molecules, but dielectric breakdown, or a short circuit in an electric circuit, also occurs in a vacuum, though at very high field strengths. Here, we discuss the breakdown phenomena in air and in a vacuum.

Figure 5.7 illustrates the mechanism of dielectric breakdown in air. As the voltage between the electrodes is increased, electrons are emitted from the cathode when the electric field strength reaches a value equivalent to the work function determined by the electrode material. This is called the field emission. When the emitted electrons collide with air molecules, the air molecules split into positive and negative ions, as shown in Fig. 5.7b. Then, as shown in Fig. 5.7c, when positive ions are attracted to the cathode and collide with the electrode, the electrons that make up the cathode material gain energy and are emitted. This is called the secondary electron emission, and these electrons further ionize the air molecules. If the number of electrons generated increases over time, the current between the electrodes may increase accordingly, leading to dielectric breakdown.

Figure 5.8 shows the relationship between the gap d [m] between the electrodes in air and the breakdown voltage V_s [V]. The dashed line in the small gap region and the solid line in the larger gap region are known as the Paschen curve. In

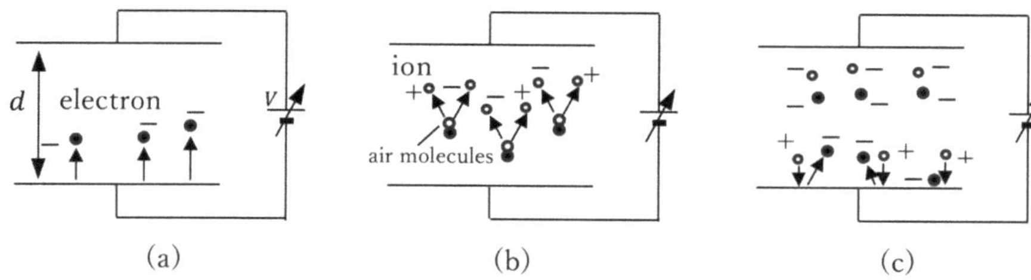

Fig. 5.7 Electrical breakdown of air due to field emission. (**a**) Field emission. (**b**) Ionized air molecules. (**c**) Secondary electron emission

Fig. 5.8 Breakdown voltage (air)

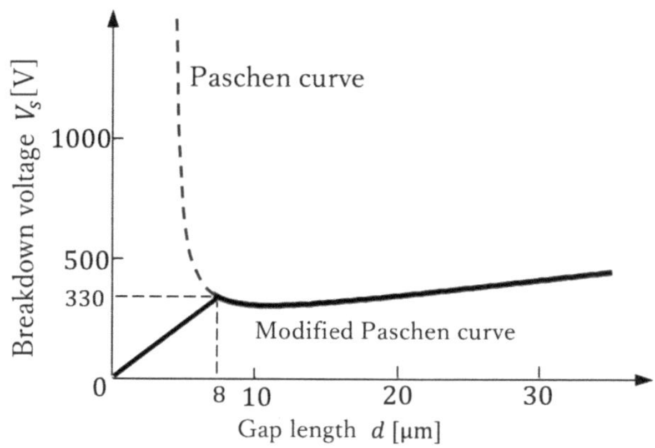

contrast, various modifications of the curve have been proposed for the gap length region $d < 8\,\mu m$, and the solid line shows a typical modified Paschen curve. In the original Paschen curve, the breakdown voltage significantly increases as shown by the dashed line in the smallest gap length range. But, in the modified Paschen curve, the voltage is proportional to the gap d in the small gap region $d < 8\,\mu m$. At $d \cong 8\,\mu m$, V_s has a value of about 330 [V], and for electrode distances above this value, the relationship between V_s and d asymptotically approaches a straight line. In the large gap range, the slope of the asymptote, i.e., the electric field intensity at breakdown, is $E_s = 3 \times 10^6$ V/m (= 3 kV/mm).

In a vacuum, on the other hand, there is no gas as a source of ions, so the above mechanism does not work. In a vacuum, two types of particles on the electrode surface are said to be responsible: one is contaminants on the surface, and the other is field-emitted electrons. First, there are microscopic clumps on the electrode surface (Fig. 5.9), making the electric field charge the clumps. The charged clumps are then driven toward the counter electrode, and when they strike the electrode, the temperature of the electrode material increases, causing the electrode to melt and vaporize. The metallic vapor is released toward the gap and forms an electric current to the electrode, resulting in dielectric breakdown. This mechanism is called

Fig. 5.9 Electrical breakdown in vacuum

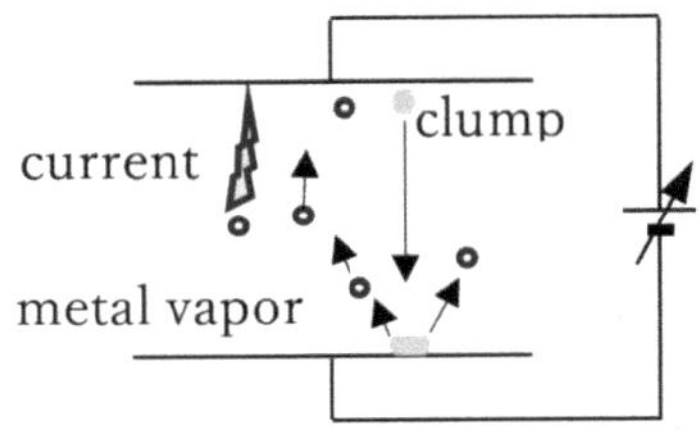

Fig. 5.10 Breakdown voltage (vacuum)

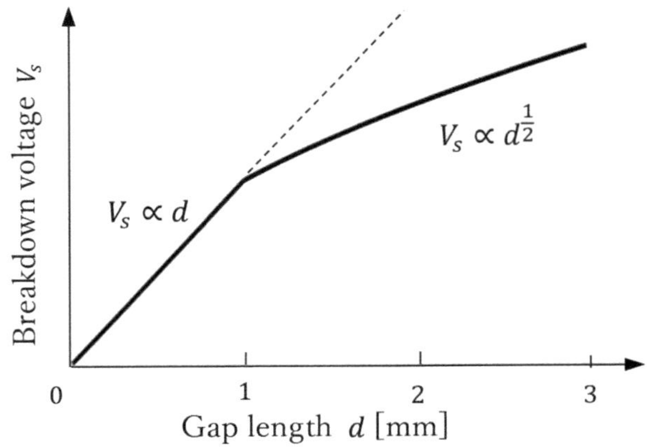

the clump theory, which explains dielectric breakdown in a vacuum, and is said to be applicable to relatively large gaps of 1 mm or more. In this case, the breakdown voltage V_s is proportional to the gap length d to the power of 1/2, and is expressed by the following equation.

$$V_s \propto d^{\frac{1}{2}} \tag{5.6}$$

In the case of a gap of 1 mm or less, when electrons reach the anode due to field emission at the cathode, the temperature of the anode rises and the gases incorporated in the electrode are released, causing the anode material to melt and vaporize, releasing its metal vapor. At the cathode, an increase in the amount of emitted electrons, or current, causes melting and vaporization of the electrode material, resulting in the release of metal vapor. This is known as the field electron emission theory. In this case, the breakdown voltage V_s is proportional to the gap length d, being expressed as

$$V_s \propto d \tag{5.7}$$

Accordingly, the electric breakdown characteristics in vacuum is shown in Fig. 5.10.

Bibliography

1. T. Sakamoto, *Electromechanical Dynamics and Control of Electric Machines* (in Japanese) (Morikita Publishing Company, 2007)
2. A. Wallash, L. Levit, Electrical breakdown and ESD phenomena for devices with nanometer-to-micron gaps, in *Proc. SPIE* (2003), p. 87
3. S. Kobayashi, Electrical breakdown phenomena in vacuum (in Japanese). J. Vac. Soc. Jpn. **56**(12), 477 (2013)

Chapter 6
Magnetic Levitation

Abstract The chapter provides a detailed physical interpretation of the forces generated by magnetic fields. Magnetic levitation is largely classified into two types: the electromagnetic suspension (EMS) and the electrodynamic suspension (EDS). Formulations are described for the EMS first, constructing a control system to establish stable levitation, which does not rely on control theory. For the EDS, two typical types of structure are described: (1) laying conducting sheet and (2) laying short-circuited coils on the ground. Analytical formulations of electromagnetic field are presented in detail for both types. Discussions are given to clarify the distribution of electromagnetic force distributions and eddy currents on the conducting sheets using numerical examples. The equations of levitation/drag ratio are derived for these structures, providing the influence of design parameters to compare the different structures. Unlike superconducting magnets on board, which are used in large-scale vehicles, the Halbach array of permanent magnets is not only inexpensive, but also has the advantage of providing a stronger magnetic field on one side where a magnetic field is needed, and almost eliminating the magnetic field on the other side. A novel analytical method for solving boundary value problems is given for this system and gives numerical results.

Compared to electrostatic levitation, magnetic levitation is characterized by its ability to generate a large force that can be applied to transportation systems, etc., without the limitations such as dielectric breakdown that can occur in electrostatic levitation. As shown in Fig. 6.1, magnetic levitation is classified into three types: electromagnetic suspension system (EMS), electrodynamic suspension system (EDS), and levitation utilizing the flux pinning unique to high-temperature superconductors. The EMS uses electromagnets attract ferromagnetic materials, while the EDS utilizes the repulsive force acting between the magnets and electric currents generated in the conductor by running the magnets near the conductor. Superconductors have Meissner effect, which excludes the magnetic flux, and the pinning force, which traps the magnetic flux due to defects in the material. Since the Meissner effect is a phenomenon that appears in a low magnetic field range, the repulsive force generated is comparatively very small.

T. Sakamoto, *Fundamentals of Producing Lift Forces*,
https://doi.org/10.1007/978-981-92-2222-3_6

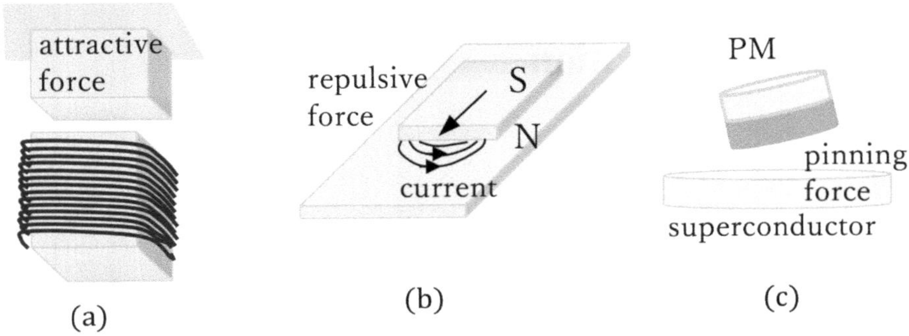

Fig. 6.1 Principles of magnetic levitation. (**a**) EMS. (**b**) EDS. (**c**) Pinning force

Note that if the repulsive force between permanent magnets is used, mechanical stability cannot be obtained according to Earnshaw's theorem mentioned in the case of electrostatic levitation. Therefore, if a mechanical device such as a wheel that creates a stabilizing force is added to stabilize the system, it would be much more reasonable to construct a system supported only by mechanical wheels in terms of the cost and handling of permanent magnets. And, thus it is not reasonable to construct a magnetic levitation system using the forces between permanent magnets.

6.1 Forces due to Magnetic Field

When considering the phenomenon of the force created by magnetic field, several different interpretations can lead to a deeper understanding, so we first discuss some basic ways of interpreting the generation of electromagnetic forces.

6.1.1 Forces Produced by Interactions of Current and Magnetic Field

The magnetic field is defined by the force exerted on a current, and if a force acts on a current, there exists magnetic field. If the current is represented by the vector $\boldsymbol{I}$ [A], the force $\boldsymbol{F}$ per unit length of the current can be expressed as $\boldsymbol{F} = \boldsymbol{I} \times \boldsymbol{B}$ [N/m] using the outer product of vectors. The relationship between the vector directions is well known from Fleming's left-hand rule. If the current density $\boldsymbol{j}$ [A/m^2] is used instead of the current $\boldsymbol{I}$ [A], the force $\boldsymbol{f}$ [N/m^3] is expressed as follows.

$$\boldsymbol{f} = \boldsymbol{j} \times \boldsymbol{B} \tag{6.1}$$

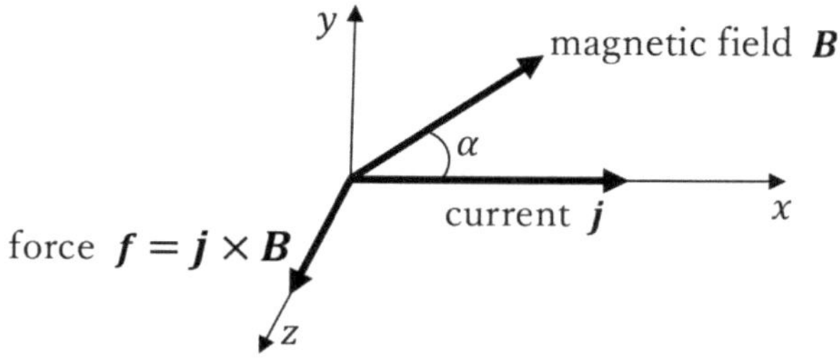

Fig. 6.2 Force generated by magnetic field and current

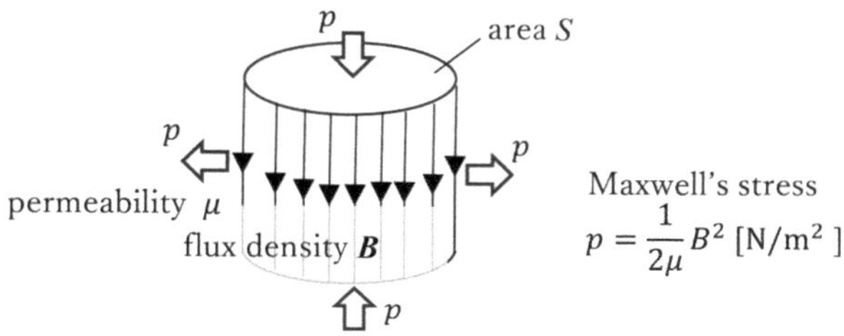

Fig. 6.3 Stress of magnetic tube of force

Assume that the current density $\boldsymbol{j}$ and magnetic field $\boldsymbol{B}$ have an angular difference α as shown in Fig. 6.2, and these two vectors correspond to the xy plane. Then, the resulting force $\boldsymbol{f}$ is in the z-axis direction perpendicular to the xy plane, which is the direction in which a right-hand thread moves when turned from the x-axis to the y-axis.

6.1.2 *Stress Transmitted through Surface*

As in the case of the electric field described in the previous chapter, we discuss how to quantitatively express the force described by Maxwell's stress (Fig. 6.3). If we consider that electromagnetic force is a stress transmitted through space via a surface, the electromagnetic force acting on an object is expressed using a mathematical expression called a stress tensor. There are two variables for the magnetic field: the magnetic flux density B and the magnetic field intensity H. The magnetic flux density B expresses the strength of the field, while H is an auxiliary variable that is necessary to express an object that exhibits magnetic properties. In the diagram, the stress is expressed in the form $p = B^2/2\mu$ [N/m^2], using the magnetic permeability μ of the space under consideration. As in the case of an electric field, for a tube created by lines of magnetic flux, which are lines of force of magnetic flux density B, a stress of $p = B^2/2\mu$ [N/m^2] acts in the longitudinal direction that tends to shrink the tube, and the stress of $p = B^2/2\mu$ [N/m^2] acts in the lateral direction that tends to expand the tube. The tube we introduced is called the tube of magnetic flux.

6.1.3 Replacing Current Loop with Magnetic Shell

When considering the magnetic field distribution created by a current loop, we find out that there exists a magnetic shell that creates the same magnetic field distribution externally. Figure 6.4 shows a current loop and the equivalent magnetic shell with the same external distribution of magnetic field B. For example, when looking at the force between two current loops as shown in Fig. 6.5a, it is complicated to use Eq. (6.1) or Fleming's left-hand rule, but it can be easily understood by replacing it with an equivalent magnetic shell as shown in Fig. 6.5b. By this replacement, it is quite easy to understand the property that a repulsive force occurs when magnetic poles of the same polarity are faced, and an attractive force occurs when magnetic poles of opposite polarity are faced.

It is also important to note that there is a qualitative relationship between current filaments similar to the relationship between magnetic shells, as shown in Fig. 6.6. That is, as shown in the diagram, currents in the same direction create an attractive force, while currents in opposite directions create a repulsive force. Using this, it is easy to see that an attractive force acts between the current loops in the case of Fig. 6.5a. It is important to look at the phenomenon from a different perspective.

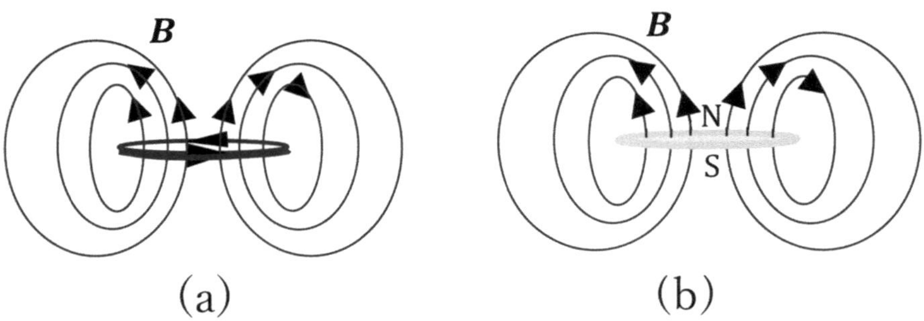

Fig. 6.4 Two sources of B. (**a**) Current loop. (**b**) Equivalent magnetic shell

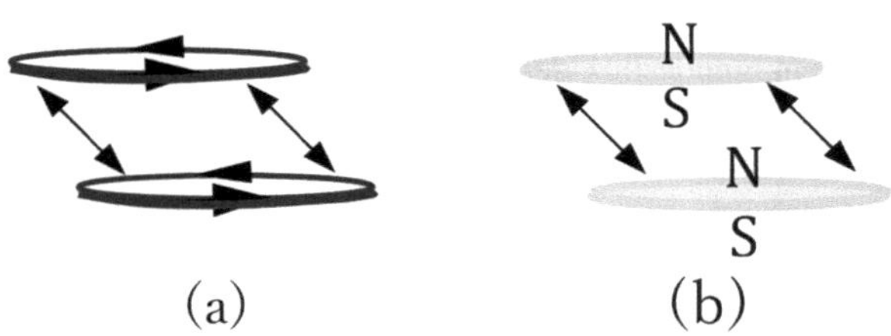

Fig. 6.5 Replacing current loop with magnetic shell. (**a**) Current loop. (**b**) Equivalent magnetic shell

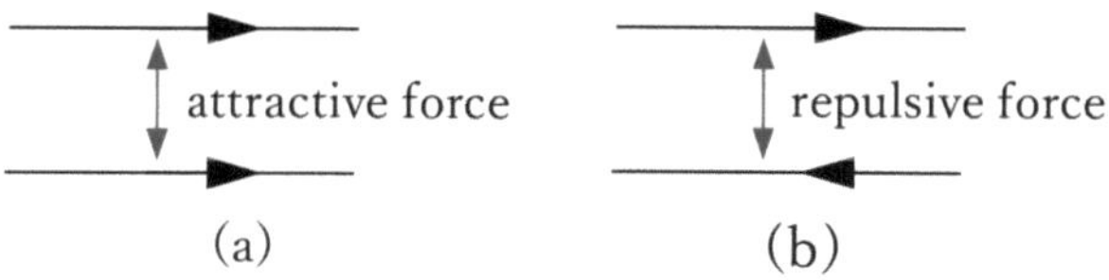

Fig. 6.6 Force between current filaments. (**a**) Same direction of currents. (**b**) Opposite direction of currents

6.2 Electromagnetic Suspension

Figure 6.7 shows an EMS model. The ferromagnetic part with high magnetic permeability is called the iron core. The iron cores are arranged in the upper and lower sections, and a coil is wound N turns around the lower iron core, and a voltage v [V] is applied to it to form an electromagnet. An attractive force acts between the upper and lower iron cores, and the electromagnet floats in the air by controlling the voltage. However, it is readily understood that without the voltage control, the electromagnet instantly adheres to the upper iron core or fall on the ground. There is even no static stability, but rather extreme instability. Therefore, the system stabilization is required to keep the suspended object at a specified position.

When the iron core is magnetized, the core itself is added to the system as a magnetic field source. First, let us use Fig. 6.8 to describe the phenomenon of the iron core magnetization. When the voltage v [V] is applied to the coil, the current i [A] is produced, being limited by the coil's electrical resistance and inductance (Fig. 6.8a). The current creates a magnetic field, and the lower iron core is magnetized according to the strength of that magnetic field, and as a result, the lower iron core becomes a magnetic field source (Fig. 6.8b). Then, the upper iron core is magnetized by the magnetic field created by the magnetized lower iron core (Fig. 6.8c). Finally, the magnetic field of the gap is formed by the magnetic fields created by the magnetized upper and lower iron cores. The attractive force acting on the gap in this case can be seen as a force due to action at a distance between the

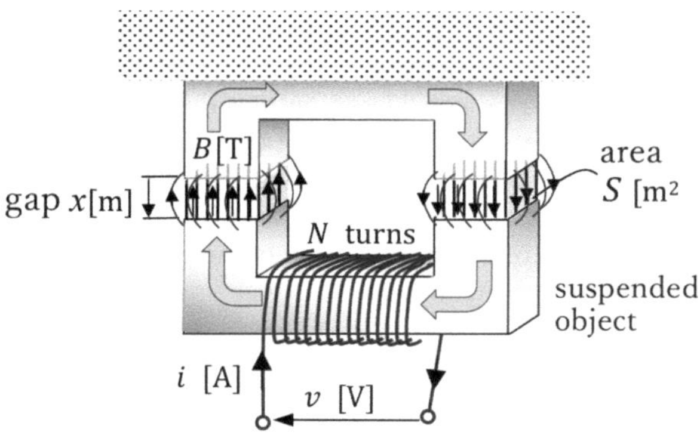

Fig. 6.7 Electromagnetic suspension

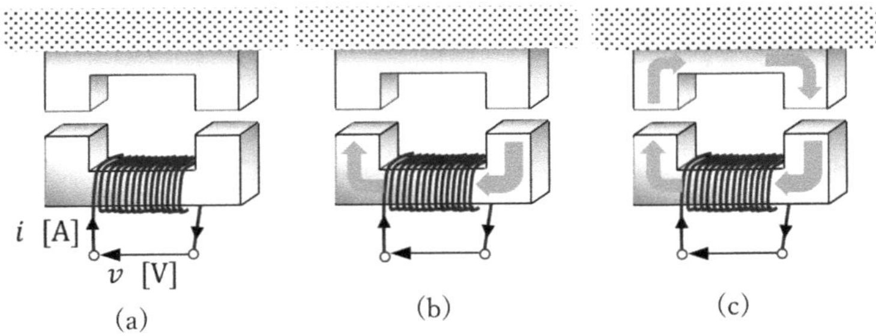

Fig. 6.8 Causality of phenomena. (**a**) Currents due to supply voltage. (**b**) Lower iron is magnetized. (**c**) Upper iron is magnetized

upper and lower iron cores that are temporarily magnetized, but it can also be seen as the local action in which the stress created by the tube of magnetic flux in the gap generates an attractive force.

6.2.1 Mathematical Representation of Controlled Object

Various control system design methods have been proposed to stabilize the EMS systems. We here, however, introduce a method for stabilizing an unstable system through physical considerations. The input given to the system for control is called the control input. In the following discussion of the EMS control systems, we use the current as the control input instead of voltage for the sake of simplicity. In other words, the controller for stabilizing the system is mathematically obtained by considering the dynamics of the electrical circuit as well as the dynamics of the EMS system, which is the target of the control system. We focus on the latter dynamics to simplify the outlook. By the way, the choice of current as the control input corresponds to the use of a current-output amplifier instead of an ordinary voltage-output power supply.

In the model shown in Fig. 6.7, Maxwell's stress due to the magnetic field acts everywhere where the magnetic field exists, but the only movable part due to the generated stress is the suspended object. In other words, the force generated in this system should be considered only at the gap. There is no time or phase delay in the generation of the force acting on the gap with respect to the current that is the source of the magnetic field. If the magnetic permeability of the iron core can be assumed to be sufficiently large, the attractive force f_a [N] generated in the gap is proportional to the inverse square of the gap length x, with k_f as the proportionality coefficient (see Fig. 6.9).

$$f_a \cong k_f \frac{i^2}{x^2} \tag{6.2}$$

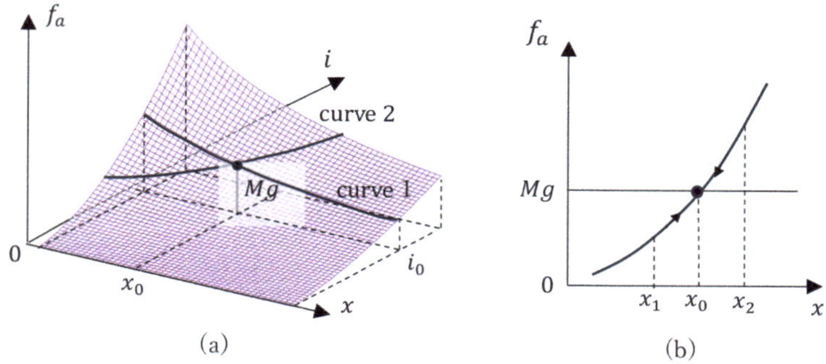

Fig. 6.9 Attractive force vs gap and current. (**a**) Attractive force. (**b**) Stabilization

If the magnetic permeability of the iron core cannot be approximated to be sufficiently large, then the gap length x is modified.

Now, assume that the suspended mass shown in Fig. 6.7 is M [kg], and the target gap is x_0 [m]. Also assume that the current required to generate the attractive force f_a of the magnitude of gravity Mg [N] is i_0 [A]. In other words, the equilibrium point of the state variables of the system is (x_0, i_0) on the (x, i) plane.

If the current is a constant value of i_0 as shown in the curve 1 in Fig. 6.9a, the attractive force decreases as the gap increases, so that the suspended object either falls or is attracted to the upper iron core. There is no restoring force to the equilibrium point, and the system is unstable. If the power supply is controlled as shown in the curve 2 by varying the current according to the gap sensor signal, the attractive force as a function of the gap is depicted in Fig. 6.9b. When the gap increases from the equilibrium point x_0 and moves to x_2, the object is pulled back to the equilibrium point x_0 because the attractive force f_a is greater than the gravitational force Mg. Conversely, if the gap decreases from the equilibrium point x_0 and moves to x_1, the attractive force f_a decreases and becomes less than the gravitational force Mg. Thus, the object is also pulled back to the equilibrium point x_0. Consequently, the system achieves static stability.

6.2.2 *Stabilizing EMS System*

Once the static stability has been achieved, the next step is to consider ensuring dynamic stability. The static restoring characteristics shown in Fig. 6.9b is simply a system in which a restoring force acts when the increment in the gap appears. If we describe K_c [N/m] as the spring constant, $\Delta x = x - x_0$ the increment in the gap, the restoring force to the equilibrium point is expressed as $K_c \Delta x$. Once a displacement Δx has occurred, the potential energy of the displacement Δx may cause persistent oscillation. Accordingly, in the next step, the attractive force must be changed to obtain a damping force based on the gap fluctuation speed signal.

To create the damping force for the motion, it is required to generate a force $\zeta_c v_x$ proportional to the time derivative of the displacement $v_x = d\Delta x/dt$, with ζ_c being the damping coefficient.

6.2.3 *Linearization within Limited Range of Variables*

As already described, it is necessary to generate the restoring force $K_c \Delta x$ proportional to Δx and the force $\zeta_c v_x$ proportional to the time derivative of the displacement $v_x = d\Delta x/dt$, where we assume that the control input is the current i in Eq. (6.2). However, because the attractive force is proportional to i^2/x^2, it may be difficult to estimate the system stability with the linear control of $K_c \Delta x$ and $\zeta_c v_x$. Therefore, we need to get the linearized formulation for the EMS system dynamics.

Let us approximate the attractive force in Eq. (6.2) expressed as a nonlinear function of the gap x and the current i, to a linear expression. Let Δx [m] be the increment from the gap x_0 of the equilibrium point, and

$$x = x_0 + \Delta x \tag{6.3}$$

Assuming that the control system can operate within a small range of the gap variation Δx, then Eq. (6.2) is linearized around the equilibrium as the desired operating point:

$$\begin{aligned} f_a(i, x) &\cong f_a(i_0, x_0) + \left.\frac{\partial f_a}{\partial i}\right|_{i_0, x_0} \Delta i + \left.\frac{\partial f_a}{\partial x}\right|_{i_0, x_0} \Delta x \\ &= k_f \frac{i_0^2}{x_0^2} + 2k_f \frac{i_0}{x_0^2} \Delta i - 2k_f \frac{i_0^2}{x_0^3} \Delta x \\ &= k_f \frac{i_0^2}{x_0^2} + k_{f1} \Delta i - k_{f2} \Delta x \end{aligned} \tag{6.4}$$

where

$$k_{f1} = 2k_f \frac{i_0}{x_0^2}, \quad k_{f2} = 2k_f \frac{i_0^2}{x_0^3}$$

Figure 6.10 shows the linearization concept. Assuming the control system works sufficiently well, the state could be kept within the expected range that Eq. (6.4) agrees well with the true behaviors. If the state variables did not fall within the desired range under an expected disturbance by the designed control system as a result of numerical simulations, it would be necessary to redesign the system to further improve control performance. The control system design should be repeated until the target performance is achieved. The first term on the right-hand side, which

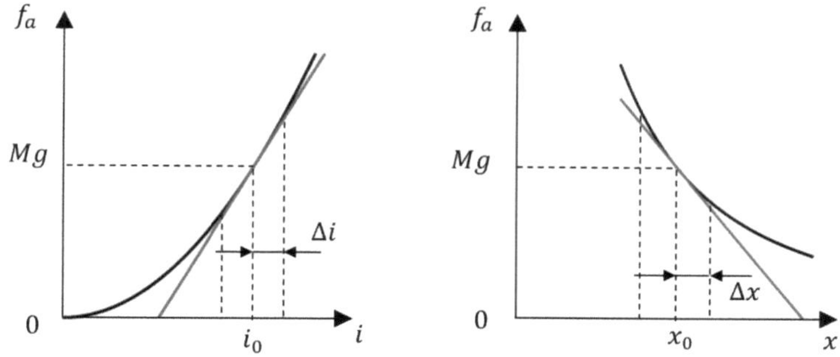

Fig. 6.10 Attractive force vs gap and current

represents the attractive force at the equilibrium point, is equal to the gravity. And we have

$$k_f \frac{i_0^2}{x_0^2} = Mg \tag{6.5}$$

The attractive force f_a is generated without any time delay relative to the current i, as described in Eq. (6.2). Let us consider how to give the change in current Δi for the changes in the gap Δx and the gap speed $d\Delta x/dt$ in order to stabilize the system.

(1) When the gap increases, i.e., $\Delta x > 0$, the attractive force must be increased to generate the restoring force, i.e., the current must be increased ($\Delta i > 0$) to decrease the gap. And in the opposite case, the current must be decreased to also create the restoring force. In other words, we can see that static stability can be obtained by the following equation.

$$\Delta i \propto \Delta x$$

(2) When the gap is about to increase with time ($d\Delta x/dt > 0$), $\Delta i > 0$ must be satisfied to increase the attractive force. And when $d\Delta x/dt < 0$, $\Delta i < 0$ must be satisfied. Therefore, the dynamic stability is obtained by the following equation.

$$\Delta i \propto \frac{d\Delta x}{dt}$$

Therefore, the control strategy is given by the equation:

$$\Delta i = k_{c1}\Delta x + k_{c2}\frac{d\Delta x}{dt} \tag{6.6}$$

where k_{c1} and k_{c2} are the positive control gains.

6.2.4 Linearized Equations of Control System

Consider the control gains k_{c1} and k_{c2} in Eq. (6.6) such that the entire system is stable. The equation of motion is given by

$$M\frac{d^2x}{dt^2} = Mg - f_a$$

Substituting Eqs. (6.3) and (6.4) into this equation, we have

$$M\frac{d^2\Delta x}{dt^2} = Mg - k_f\frac{i_0^2}{x_0^2} - k_{f1}\Delta i + k_{f2}\Delta x \tag{6.7}$$

From Eq. (6.5),

$$\frac{d^2\Delta x}{dt^2} = -\frac{k_{f1}}{M}\Delta i + \frac{k_{f2}}{M}\Delta x \tag{6.8}$$

Now, if we employ the control strategy of Eq. (6.6), then we have the motion of equation for the entire system:

$$\frac{d^2\Delta x}{dt^2} + \frac{k_{c2}k_{f1}}{M}\frac{d\Delta x}{dt} + \frac{k_{c1}k_{f1} - k_{f2}}{M}\Delta x = 0 \tag{6.9}$$

The resulting equation has the same form as the equation of motion for a quadratic system shown in Fig. 6.11, which provides well-known characteristics in terms of the parameters. In the diagram, the parameter ζ is the damping ratio that is determined by the damper, and ω_n the natural angular frequency that is determined by the spring constant. Therefore, the damping ratio ζ determines the damping of the vibration, and ω_n leads to the speed of restoration to the equilibrium point, which represents the response speed in the control system.

In Eq. (6.9), any displacement Δx from the equilibrium point will cause oscillations according to the dynamics of this differential equation.Therefore, the spring and damping coefficients of the levitation system must be designed so that the system is stable and has the desired motion.

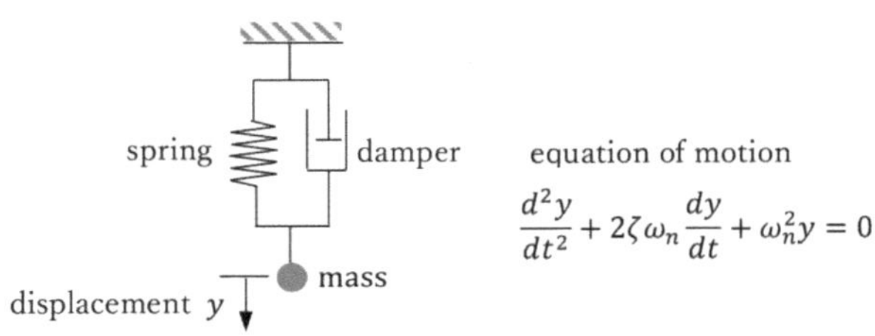

Fig. 6.11 Mechanical model of second order system

6.2.5 Controller Gains

The standard form of a second-order vibration system with respect to displacement $y(t)$ is given by the following equation, where ω_n [rad/s] is the natural angular frequency and ζ is the damping ratio, as was already shown in Fig. 6.11.

$$\frac{d^2 y}{dt^2} + 2\zeta\omega_n \frac{dy}{dt} + \omega^2 y = 0 \tag{6.10}$$

ω_n indicates the speed of response; the larger the value, the faster the response. When the damping ratio ζ has a positive value, the system is stable and has the following properties.

- $0 < \zeta < 1$ The system oscillates about the equilibrium and is slow to decay to the equilibrium.
- $\zeta \geq 1$ The system decays to the equilibrium without oscillating.

The larger the damping ratio ζ, the higher the stability, but since the control performance deteriorates instead, a value of about $1/\sqrt{2}$ may be appropriate. Furthermore, Eq. (6.8) represents the dynamics of the suspended object assuming incremental changes for the current and gap. If we ignore the term associated with the current Δi on the right-hand side, it represents the movement when the current is constant, which is expressed by the following equation.

$$M\frac{d^2\Delta x}{dt^2} - k_{f2}\Delta x = 0 \tag{6.11}$$

This equation describes the motion for a system in which mass M is subjected to a force by a negative spring constant $-k_{f2}$. If we use $\omega_M = \sqrt{k_{f2}/M}$, the solution is expressed as the linear sum of the functions $e^{\omega_M t}$ and $e^{-\omega_M t}$. Therefore, if the control input is delayed, the suspended object may either fall on the ground or sticks to the iron core at a speed characterized by $\omega_M = \sqrt{k_{f2}/M}$ as shown in Fig. 6.12.

Now, by comparing Eqs. (6.10) and (6.9), we obtain the following relationships:

$$\begin{aligned} \omega_n^2 &= \frac{k_{c1}k_{f1} - k_{f2}}{M} \\ \zeta &= \frac{k_{c2}k_{f1}}{2\omega_n M} \end{aligned} \tag{6.12}$$

The former of Eq. (6.12) is the natural angular frequency ω_n to the power of 2. Then, firstly the right-hand side should be positive, i.e., $k_{c1}k_{f1} - k_{f2} > 0$. And, the control gain must satisfy at least $k_{c1} > k_{f2}/k_{f1}$. Consequently, if we specify the natural angular frequency ω_n, the control gain k_{c1} is determined as

$$k_{c1} = \frac{M\omega_n^2 + k_{f2}}{k_{f1}} \tag{6.13}$$

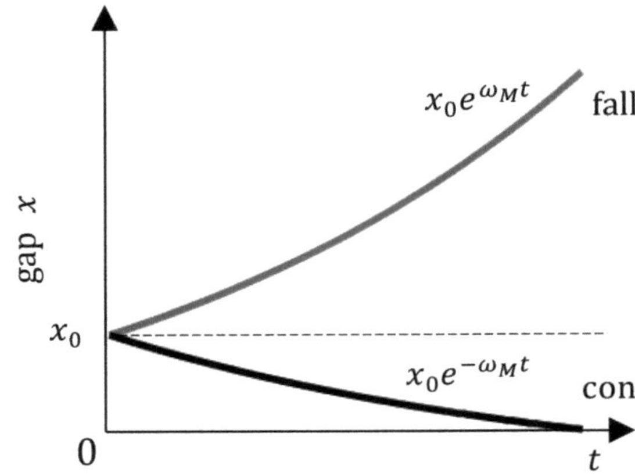

Fig. 6.12 Instability of EMS (without control)

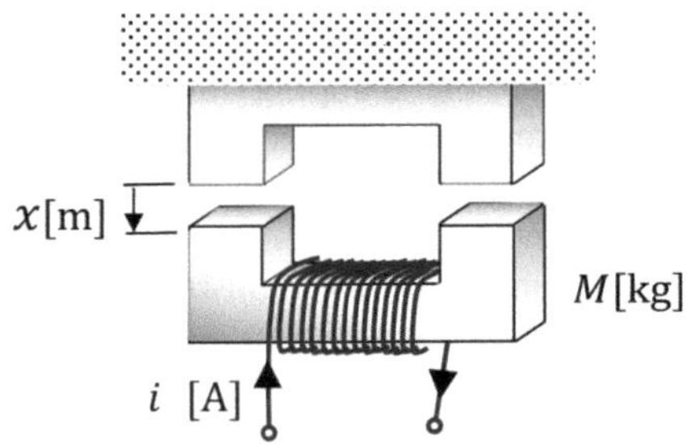

Fig. 6.13 EMS control system

And, the controller gain k_{c2} is given by specifying the damping ratio ζ in the latter of Eq. (6.12):

$$k_{c2} = \frac{2\zeta\omega_n M}{k_{f1}} \tag{6.14}$$

Example 6.1 (Control System Design of an EMS System) In the EMS system shown in Fig. 6.13, the mass of the suspended object is $M = 0.925$ kg, the attraction force coefficient k_f is 6.25×10^{-5} Nm^2/A^2, and the reference gap length x_0 is 5 mm. Apply the control system design methodology we showed above to this system.

Solution From Eq. (6.5), the steady state current is

$$i_0 = x_0\sqrt{\frac{Mg}{k_f}} = 0.005 \times \sqrt{\frac{0.925g}{6.25 \times 10^{-5}}} = 1.904\,[\text{A}]$$

The coefficients k_{f1} and k_{f2} of Eq. (6.4) are then given as

$$k_{f1} = 2k_f\frac{i_0}{x_0^2} = 2 \times 6.25 \times 10^{-5} \times \frac{1.904}{0.005^2} = 0.952$$

$$k_{f2} = 2k_f\frac{i_0^2}{x_0^3} = 2 \times 6.25 \times 10^{-5} \times \frac{1.904^2}{0.005^3} = 363$$

In the step response of a second-order system, its envelope is given by $e^{-\zeta\omega_n t}$. Then, regarding the settling time T_s [s] that falls within ±5% of the steady-state output, the following equation holds.

$$e^{-\zeta\omega_n T_s} = 0.05$$

Thus, we have

$$T_s \cong \frac{3}{\zeta\omega_n}$$

If the natural angular frequency is specified as $\omega_n = 30$ rad/s, and the damping factor as $\zeta = 1/\sqrt{2} \cong 0.7$, the settling time becomes $T_s \cong 0.14$ s, which may be a reasonable performance. In this case, the control gains are calculated as follows:

$$k_{c1} = \frac{M\omega_n^2 + k_{f2}}{k_{f1}} = \frac{0.925 \times 30^2 + 363}{0.952} = 1256$$

$$k_{c2} = \frac{2\zeta\omega_n M}{k_{f1}} = \frac{2 \times 0.7 \times 30 \times 0.925}{0.952} = 40.8$$

In general, the natural angular frequency in a mechanical system is determined by the spring constant and the mass. That is, specifying a higher natural angular frequency ω_n translates into a higher spring constant created by the control system. However, specifying the spring constant of the control system determines the attractive force for a gap displacement Δx, i.e., the change in the current Δi is required proportional to the displacement (see Eq. (6.6)). Thus, we need to be careful about specifying the appropriate value of the natural angular frequency, because an excessively large natural angular frequency ω_n may require an excessively large control current Δi from the amplifier.

By assuming that a gap displacement Δx has occurred, let us check whether the required current is appropriate from the perspective of amplifier cost, in comparison with the steady-state current. If we ignore the gap displacement speed $d\Delta x/dt$ in Eq. (6.6), and assume a gap displacement of 1 mm, for example,

$$\Delta i = k_{c1}\Delta x = 1256 \times 0.001 = 1.256\,\text{A}$$

The calculated value of the current increase is about the same magnitude as the steady-state current $i_0 = 1.904$ A, which may be within the acceptable range. ◇

6.2.6 *Control System Simulations*

Now that the gains of the control system have been determined, we confirm the control system performance with a simulation software to verify the validity of our design results. It is then necessary to represent the mathematical contents of the

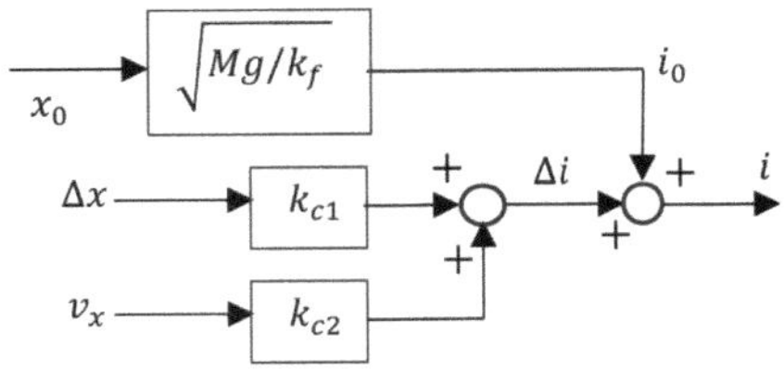

Fig. 6.14 Current reference

designed system on a computer through a block diagram. To show this procedures, each of the mathematical expressions obtained above is represented in a block diagram, and finally they are integrated to complete the diagram.

The coil current i is composed of the current i_0 required steadily to maintain the gap at the equilibrium point, and the current Δi given to perform the restoring action when the gap deviates from the equilibrium point. This relation is expressed in a block diagram shown in Fig. 6.14, where we used $v_x = d\Delta x/dt$ and the relationship in Eq. (6.5). The circle in the diagram is the symbol for addition and subtraction, called the summing point. The parts with coefficients written in the squares indicate that the variable that is reached is multiplied and output, and this coefficient is usually called the gain. The input signals in the diagram are the reference gap x_0 [m], the gap displacement Δx [m], and the gap displacement speed v_x [m/s], which are multiplied by the gain and added at the summing point. And, the current command value is given to the current amplifier, where it is assumed in the diagram that the amplifier outputs the same value of command. If the response of the current amplifier were not sufficient, then the current amplifier would have to be replaced with a voltage amplifier, leading to redesigning the control system with the voltage input. The current i_0 corresponding to the reference gap x_0 is constructed using the relationship $i_0 = \sqrt{Mg/k_f} \cdot x_0$ of Eq. (6.5).

Next, from Eq. (6.8) we get

$$x = \frac{1}{M} \int \left\{ \int (Mg - f_a) dt \right\} dt \tag{6.15}$$

It is important to note here that control system simulation software must configure blocks in accordance with the law of causality, where a gap displacement occurs when an input of $Mg - f_a$ is given. That is, $Mg - f_a$ is the cause, and by integrating this twice and multiplying it by $1/M$, the resulting gap x is obtained.

In control engineering, this mathematical relationship is expressed by using a technique called the Laplace transform, where the function of time t is transformed to a function of the complex variable s. As a result, the integral operation with respect to time t is expressed by multiplying by $1/s$. Therefore, when Eq. (6.15) is shown in Fig. 6.15. The output of the block $1/Ms$ is the gap velocity dx/dt, and when this is passed through the block $1/s$, the gap x is obtained. When the Laplace transform is used, differential and integral operations with respect to time are expressed in algebraic operations in terms of the variables s. The part of the function surrounded by square symbols in the diagram is called the transfer function.

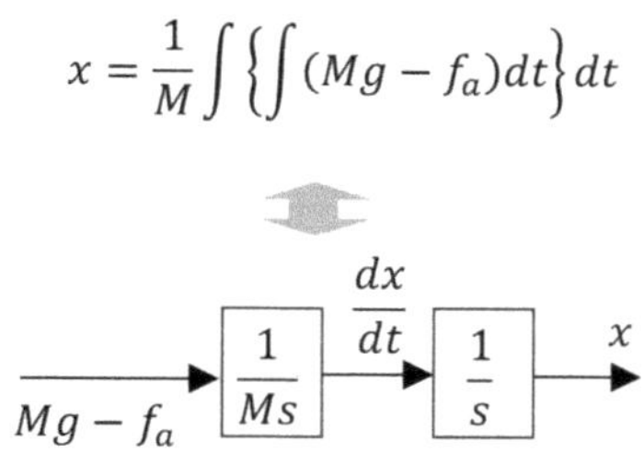

Fig. 6.15 Integrator 1/s

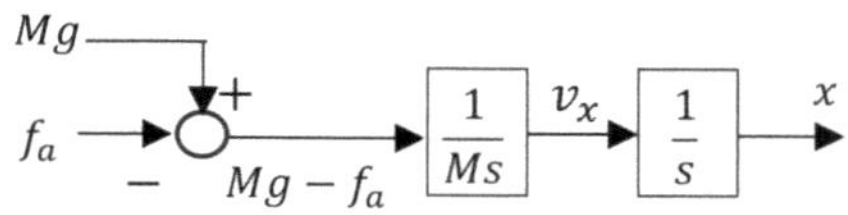

Fig. 6.16 Block diagram of mechanical dynamics

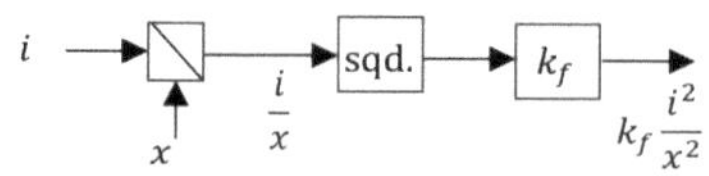

Fig. 6.17 Attractive force

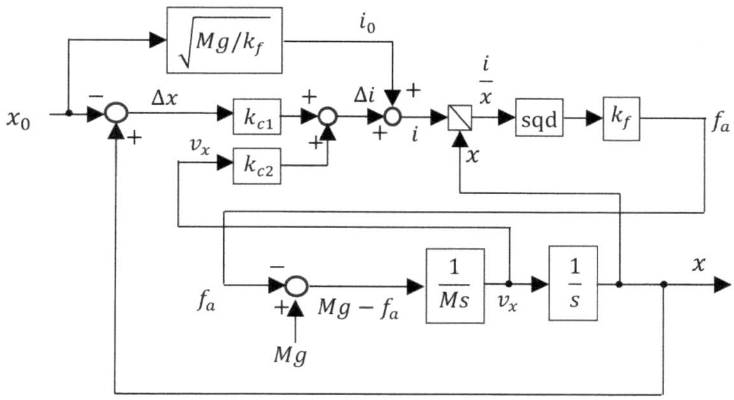

Fig. 6.18 Block diagram of EMS system

If we express the equation $Mg - f_a$ in Fig. 6.16 using the summing points, we get Fig. 6.17. And if we get Fig. 6.17 from Eq. 6.2, where the division of two variables is represented by a square with a line through it, and the square is represented by "sqd.".

Combining the above block diagrams into one, we obtain Fig. 6.18. Note that the representation of block diagrams may differ slightly depending on the software. After the block diagram is completed, the simulation is carried out. In the simulation, when the reference gap x_0, which represents the steady-state gap, is given, the error signal $\Delta x(t)$ is given to the control system, and the control current $\Delta i(t)$ is generated. The steady-state current i_0 required to maintain the reference gap is added, and $i(t)$ is given to the current amplifier as the command value. Meanwhile, the attractive force $f_a(t)$ is calculated based on the current $i(t)$ and the value of the gap $x(t)$, the gravity Mg is given as a disturbance, and the gap value $x(t)$ is produced.

Example 6.2 (Control System Simulation) Figure 6.19 shows the results of a simulation in which the initial gap value is set to $x(0) = 7$ mm and the reference gap is $x_0 = 5$ mm. Since the mass of the suspended object is $M = 0.925$ kg, the magnetic force required in the steady state is 9.07 N. When a difference Δx occurs between the reference gap value x_0 and the actual gap x, the control current Δi is generated, as can be seen from the block diagram in Fig. 6.17. Therefore, since there is a deviation of +2 mm at $t = 0$, in order to bring the gap closer to the reference of 5 mm, $\Delta i(0) = 2.51$ A is generated as shown in Fig. 6.19a, and a large magnetic force of $f_a(0) = 24.9$ N is generated. As a result, it can be confirmed that the equilibrium point has been almost completely settled after about 0.15 s just as designed. ◇

6.3 EDS System Using Conducting Sheet

Consider a vehicle model with an alternating-polarity magnet array moves above a conducting sheet, as shown in Fig. 6.20. When viewed from the conducting sheet, the magnetic poles appear to alternate, and the magnetic field created by the magnets fluctuates over time, creating an electric field due to the temporal fluctuations in the magnetic field. Since the conducting sheet is placed in a location where an electric field is generated, eddy currents flow in the conducting sheet. These eddy currents and the magnetic field created by the onboard magnets create a repulsive force between them, and this phenomenon is used to obtain a levitation force, which is the principle of the EDS.

Because the EDS requires a sufficiently strong magnetic field to create repulsive forces, superconducting coil can be the only candidate for passenger transport vehicles, while a permanent magnet arrangement called the Halbach array, which maximizes the generated magnetic field, is a candidate for small-scale vehicles. With the EMS method described in the previous section, if a large gap is secured for safety, the leakage flux increases, resulting in an extreme drop in efficiency. On the other hand, with the EDS, the efficiency is not an issue since the power supply is not necessary other than the cooling system for superconducting coils.

In this section, we introduce the methodology and concepts of electromagnetic field analysis of the eddy current problems, which is not particularly easy to handle, and provide a quantitative discussion of EDS levitation.

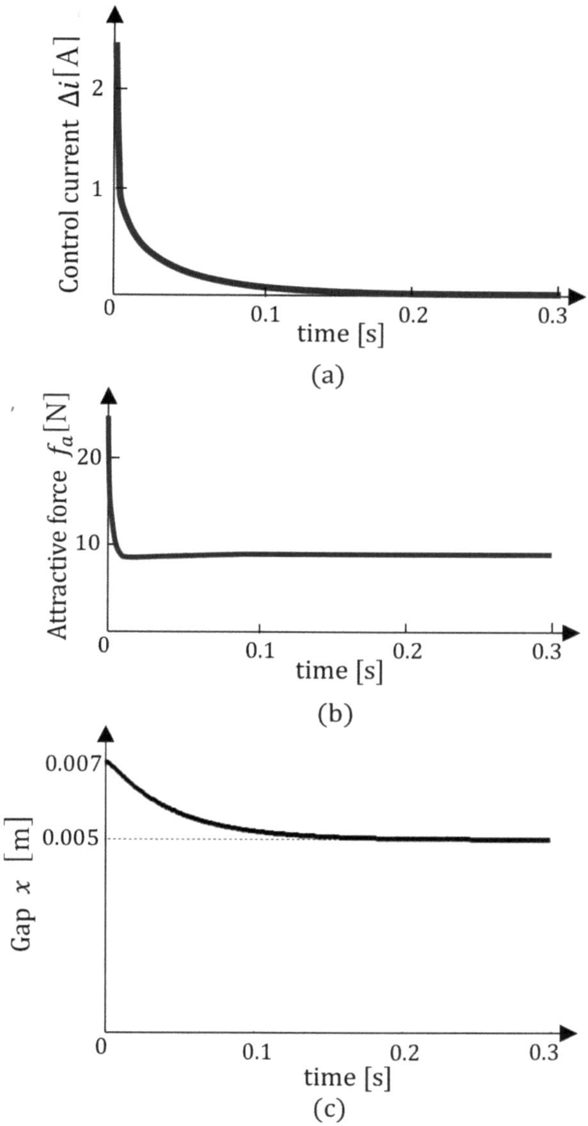

Fig. 6.19 Control of EMS system. (**a**) Control current. (**b**) Attractive force. (**c**) Gap

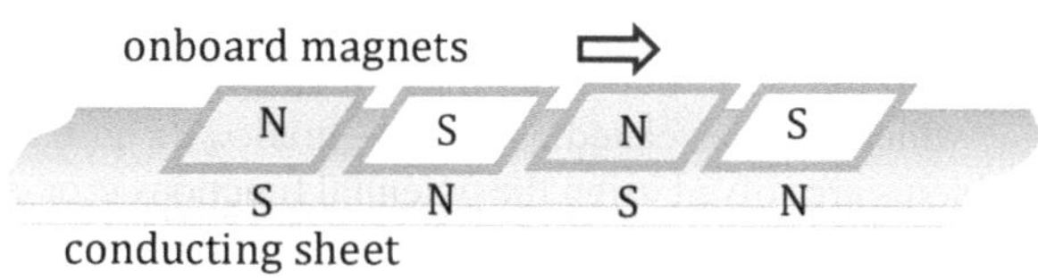

Fig. 6.20 Principle of EDS

phenomenon : electric charge ρ produces electric field $\boldsymbol{E}$
mathematical problem : solve governing equation of electric scalar potential φ ➡ $\boldsymbol{E} = -\mathrm{grad}\varphi$

(a)

phenomenon : magnetic charge ρ_m produces magnetic field $\boldsymbol{H}$
mathematical problem : solve governing equation of magnetic scalar potential φ_m ➡ $\boldsymbol{H} = -\mathrm{grad}\varphi_m$

(b)

Fig. 6.21 Approach to electrostatic and magnetostatic problems. (**a**) Electrostatic field problem. (**b**) Magnetostatic field problem

6.3.1 Approach to Electromagnetic Analysis

Figure 6.21 shows a typical approach for electrostatic and magnetic field analyses. It is well known that when solving problems in electrostatic fields, instead of solving equations for the electric field intensity $\boldsymbol{E}$ constituted of three dimensional components in (x, y, z), it is reasonable to solve the electrostatic potential or the electric scalar potential φ, which is related as $\boldsymbol{E} = -\mathrm{grad}\varphi$.

There is redundancy in treating the three components of the electric field $\boldsymbol{E}$ as unknown variables. If the charge density distribution $\rho(x, y, z)$ is given in the air, the governing equation $\mathrm{div}\boldsymbol{E} = \rho/\varepsilon_0$ holds, where ε_0 is the dielectric constant. But in a three-dimensional problem, the electric field strength $\boldsymbol{E}$ has three variable, while there is only one equation. Therefore, to remove the redundancy of this variable, we should introduce the electrostatic potential φ to get $\boldsymbol{E} = -\mathrm{grad}\varphi$.

Then, we obtain $\nabla^2 \varphi = -\rho/\varepsilon_0$, and the number of unknown variables becomes equal to the number of equations. Note that the operator grad is $(\partial/\partial x)\boldsymbol{e}_x + (\partial/\partial y)\boldsymbol{e}_y + (\partial/\partial z)\boldsymbol{e}_z$, which calculates the gradient in three dimensions, so that the electric field is expressed by $\boldsymbol{E} = -\mathrm{grad}\varphi = (\partial\varphi/\partial x)\boldsymbol{e}_x + (\partial\varphi/\partial y)\boldsymbol{e}_y + (\partial\varphi/\partial z)\boldsymbol{e}_z$. The operator div is called the divergence operator, and this operation calculates the degree of "springing out" of an electric field in the coordinates of interest. In other words, if $\mathrm{div}\boldsymbol{E} > 0$ at a certain point, it means that the electric field $\boldsymbol{E}$ "springs out" from that point, while $\mathrm{div}\boldsymbol{E} < 0$ translates into that the electric field "terminates" at that point.

Static magnetic fields can thus be analyzed using the magnetic scalar potential φ_m, which can be related as $\boldsymbol{H} = -\mathrm{grad}\varphi_m$ (Fig. 6.21b). Accordingly, associated problems are solved using the potential functions φ or φ_m, rather than using physical variables $\boldsymbol{E}$ or $\boldsymbol{H}$.

phenomenon : current $\boldsymbol{j}$ and electric charge ρ produces $(\boldsymbol{E}, \boldsymbol{B})$
mathematical problem : solve governing equations of magnetic vector potential $\boldsymbol{A}$ and electric scalar potential φ ➡ $\boldsymbol{E} = -\frac{\partial \boldsymbol{A}}{\partial t} - \mathrm{grad}\varphi$, $\boldsymbol{B} = \mathrm{rot}\boldsymbol{A}$

Fig. 6.22 Approach to general electromagnetic problem

In general electromagnetic field problems, the formulations become somewhat complicated because the electromagnetic field fluctuates over time. Electric currents generate a magnetic flux density $\boldsymbol{B}$, where the divergence of $\boldsymbol{B}$ always holds true as $\mathrm{div}\boldsymbol{B} = 0$. In that case, if we introduce $\boldsymbol{B} = \mathrm{rot}\boldsymbol{A}$, this equation holds identically, and vector $\boldsymbol{A}$ is called the magnetic vector potential (Fig. 6.22). Note that the operator rot is an operation that numerically determines the degree to which a vortex is formed, in exactly the same way that the degree of slope of a road can be determined by differentiating the height of the road with respect to the position.

In this case, using Maxwell's equation $\mathrm{rot}\boldsymbol{E} = -\partial \boldsymbol{B}/\partial t$ and the vector formula $\mathrm{rot}(\mathrm{grad}\varphi) = 0$, we obtain $\boldsymbol{E} = -\partial \boldsymbol{A}/\partial t - \mathrm{grad}\varphi$, where φ is an arbitrary scalar function. Here, if we limit the problem to the one in air where dielectrics are not considered, we can write $\mathrm{div}\boldsymbol{E} = \rho/\varepsilon_0$. Therefore, we obtain $\mathrm{div}\boldsymbol{E} = \mathrm{div}(-\partial \boldsymbol{A}/\partial t - \mathrm{grad}\varphi)$.

In order to eliminate the uncertainty of the magnetic vector potential, we here introduce the condition $\mathrm{div}\boldsymbol{A} = 0$ known as the Coulomb gauge. Then we have $\mathrm{div}(-\partial \boldsymbol{A}/\partial t - \mathrm{grad}\varphi) = -\nabla^2\varphi$, and we obtain $\nabla^2\varphi = -\rho/\varepsilon_0$, in which φ is governed only by ρ.

In the analytical approach, due to the complexity of dealing with induced charges ρ resulting from the discontinuities of conducting regions, it is necessary to complete the analysis using only the magnetic vector potential $\boldsymbol{A}$ instead of $(\boldsymbol{A}, \varphi)$. For the EDS problem described in this section, the conducting sheet is considered to extend parallel to the surface of the on-board magnets. In other words, we employ a method to ensure that there are no discontinuities in the conducting sheet in the direction of the generated electric field from the magnets.

6.3.2 Formulation of Onboard Magnets or Coils

In the analysis of EDS, we assume that the coil current array as shown in Fig. 6.23a, and has a velocity v [m/s] in the x-axis, in which the coil current alternates in the x-axis to create alternating magnetic polarities. In the y-axis direction, which is the lateral direction, we consider infinite number of coil currents that are placed in alternate polarities, which is a convenient way to consider the width of the on-board coils in trigonometric terms.

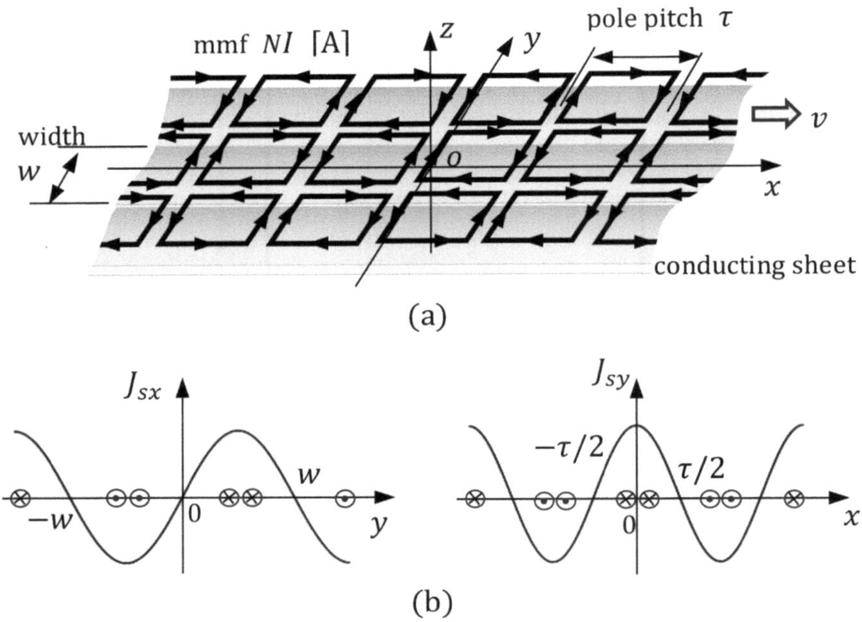

Fig. 6.23 EDS model and coil surface currents. (**a**) EDS model. (**b**) Sinusoidal approximation of onboard coil currents ($t = 0$)

The conducting sheet on the ground has the infinite length and the width of w [m] located below the on-board coil currents along the x-axis, and adjacent conducting sheets in the lateral direction are electrically connected. In other words, the conducting sheet extends infinitely in the xy plane. If we were to consider an insulated conductor plate of w [m] in the y-axis direction, we would need to consider induced charges to contain the eddy currents within this width, which would require $(\boldsymbol{A}, \varphi)$ as the electromagnetic potential. However, in this case, the infinitely long conducting sheets are electrically connected to each other laterally, and by setting up a model in which the coil currents are arranged in a half period of width w [m], it is possible to create a situation in which the eddy current distribution is contained within the width w [m]. In addition, the electric scalar potential φ is no longer necessary, and the analysis can be performed using only the magnetic vector potential $\boldsymbol{A}$.

Now, the conducting sheet has the thickness d [m] and the conductivity σ [S/m], and in the analysis the product of the conductivity and the thickness σd is kept constant while the thickness d is assumed to be infinitesimal. Using the Dirac delta function $\delta(z)$, the distribution of σd can be expressed as $\sigma d \delta(z)$. The on-board coil current is also described as a surface current to have an infinitesimal thickness.

The coil current array in Fig. 6.23a is now expressed as a traveling wave in the x-axis direction with the sinusoidal surface current $\boldsymbol{J}_s(x, y, t) = (J_{sx}, J_{sy}, 0)$ shown in Fig. 6.23b. The diagram shows that the distribution of J_{sx} is an odd function, while the distribution of J_{sy} an even function, which are expressed by the following equation.

$$J_{sx} = J_{smx} \sin \frac{\pi y}{w}, \quad J_{sy} = J_{smy} \cos \frac{\pi x}{\tau}$$

If a current I flows in each coil and the number of turns is N, the coil magnetomotive force is expressed as NI [A]. The integral value of the half cycle of the current density at time $t = 0$ in the figure is equal to the current $2NI$ of the two adjacent coil filaments, as shown by the cross and dot symbol in the diagram, i.e.,

$$2NI = \int_0^w J_{sx} dy = \frac{2w}{\pi} J_{smx} = \int_{-\tau/2}^{\tau/2} J_{sy} dx = J_{smy}$$

which gives the relation

$$J_{smx} = \frac{\pi}{w} NI, \quad J_{smy} = \frac{\pi}{\tau} NI \tag{6.16}$$

Therefore, we obtain the coil surface currents $\boldsymbol{J}_s = (J_{sx}, J_{sy}, 0)$ [A/m] as

$$\begin{aligned} J_{sx} &= -J_{smx} \sin(\omega t - kx) \sin\left(\frac{\pi}{w} y\right) \\ &= \mathrm{Re}\left\{ j J_{smx} e^{j(\omega t - kx)} \sin\left(\frac{\pi}{w} y\right)\right\} \\ &= \mathrm{Re}\{\dot{J}_{sx}(x, y)\} \end{aligned} \tag{6.17}$$

$$\begin{aligned} J_{sy} &= J_{smy} \cos(\omega t - kx) \cos\left(\frac{\pi}{w} y\right) \\ &= \mathrm{Re}\{\dot{J}_{sy}(x, y)\} \end{aligned} \tag{6.18}$$

where, "Re" indicates the operation of taking the real part, and the angular frequency ω and the wave number k are given as $\omega = \pi v/\tau$ [rad/s] and $k = \pi/\tau$ [rad/m], respectively, and

$$\begin{aligned} \dot{J}_{sx}(x, y) &= j J_{smx} e^{j(\omega t - kx)} \sin\left(\frac{\pi}{w} y\right) \\ \dot{J}_{sy}(x, y) &= J_{smy} e^{j(\omega t - kx)} \cos\left(\frac{\pi}{w} y\right) \end{aligned} \tag{6.19}$$

Now that the on-board coil currents were expressed by Eqs. (6.17) and (6.18), the next step is the analysis using the magnetic vector potential. We derive the magnetic vector potentials separately for the coil currents and eddy currents:

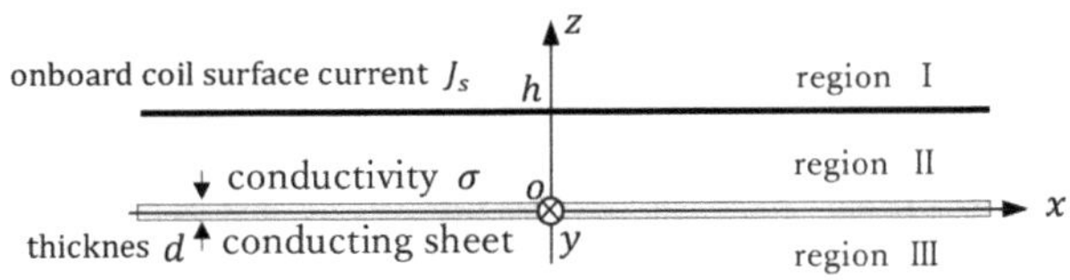

Fig. 6.24 Cross-section of model

(1) Consider the three regions shown in Fig. 6.24, set the on-board coil currents and conducting sheet as the boundary surfaces of the analysis, and include these surface currents as boundary conditions.
(2) Find the electric field $\boldsymbol{E}_s$ [V/m] in the conducting sheet caused by the on-board coil current $\boldsymbol{J}_s$ [A/m].
(3) Describe the eddy current $\boldsymbol{J}_e$ [A/m] generated in the conducting sheet using unknown coefficients.
(4) Formulate the electric field $\boldsymbol{E}_e$ [V/m] created by the eddy current $\boldsymbol{J}_e$ [A/m].
(5) Find the unknown coefficients of $\boldsymbol{J}_e$ using the relation $\boldsymbol{J}_e = \sigma d(\boldsymbol{E}_s + \boldsymbol{E}_e)$ to obtain the eddy current solution in the conducting sheet.

6.3.3 *Magnetic Vector Potential of On-Board Coil Currents*

The surface current $\boldsymbol{J}_s$ [A/m] of the on-board coils located at the coordinate $z = h$ has a sinusoidal waveform with a spatial magnetic pole pitch τ [m] in the direction of travel of the vehicle as shown in Eqs. (6.17) and (6.18), and the distribution moves in the positive direction of the x-axis at a speed v [m/s]. If the intensity of the magnetic field generated by this surface current is taken as $\boldsymbol{H}_s$ [A/m], then in free space other than the coordinate $z = h$, $\nabla \times \boldsymbol{H}_s = 0$. Defining the magnetic vector potential created by this current $\boldsymbol{A}_s$, we get

$$\nabla \times \boldsymbol{H}_s = \nabla \times \left(\frac{1}{\mu_0}\nabla \times \boldsymbol{A}_s\right) = \frac{1}{\mu_0}\nabla(\nabla \cdot \boldsymbol{A}_s) - \nabla^2 \boldsymbol{A}_s = 0$$

Again, the coil current is given as a boundary condition, so that the right hand side is set to 0. Since $\boldsymbol{A}_s$ cannot be uniquely determined in this state, if we set the Coulomb gauge $\nabla \cdot \boldsymbol{A}_s = 0$ to eliminate the uncertainty of the magnetic vector potential,

$$\nabla^2 \boldsymbol{A}_s = 0 \tag{6.20}$$

The solution in free space is obtained by solving Laplace's equation, but this equation does not include a forcing term; instead, the magnetic field source is given by the boundary conditions. Furthermore, if the currents only have x and y components, then the magnetic vector potential also have only x and y components just like the currents, because the current density $\boldsymbol{j}$ creates the magnetic vector potential $\boldsymbol{A}$ given by the general equation $\nabla^2 \boldsymbol{A} = -\mu_0 \boldsymbol{j}$. That is, the magnetic vector potential created by the on-board coil currents is written as $\boldsymbol{A}_s = (A_{sx}, A_{sy}, 0)$.

Let us derive the boundary conditions for the general case where there is a surface current $\boldsymbol{J} = (J_x, J_y, 0)$ [A/m] flowing in the xy-plane at the boundary between the region 1 and the region 2, as shown in Fig. 6.25. If $\boldsymbol{n}$ is taken as the normal unit vector to the boundary surface, and the region name is added to the superscript of variables, then the boundary conditions for the magnetic field intensities $\boldsymbol{H}^1$ and $\boldsymbol{H}^2$ at the boundary are derived from Maxwell's equation $\nabla \times \boldsymbol{H} = \boldsymbol{j}$:

$$\boldsymbol{n} \times (\boldsymbol{H}^1 - \boldsymbol{H}^2) = \boldsymbol{J}$$

Substituting the equation

$$\boldsymbol{H} = \frac{\boldsymbol{B}}{\mu_0} = \frac{1}{\mu_0} \nabla \times \boldsymbol{A}$$

we have

$$\boldsymbol{n} \times \boldsymbol{H} = \frac{1}{\mu_0} \boldsymbol{e}_z \times \nabla \times \boldsymbol{A} = -\frac{1}{\mu_0} \frac{\partial A_x}{\partial z} \boldsymbol{e}_x - \frac{1}{\mu_0} \frac{\partial A_y}{\partial z} \boldsymbol{e}_y$$

where $\boldsymbol{n} = \boldsymbol{e}_z$. Thus, we obtain the boundary conditions for the x and y components of surface current as

$$\begin{aligned} -\frac{1}{\mu_0} \frac{\partial A_x^1}{\partial z} + \frac{1}{\mu_0} \frac{\partial A_x^2}{\partial z} &= J_x \\ -\frac{1}{\mu_0} \frac{\partial A_y^1}{\partial z} + \frac{1}{\mu_0} \frac{\partial A_y^2}{\partial z} &= J_y \end{aligned} \tag{6.21}$$

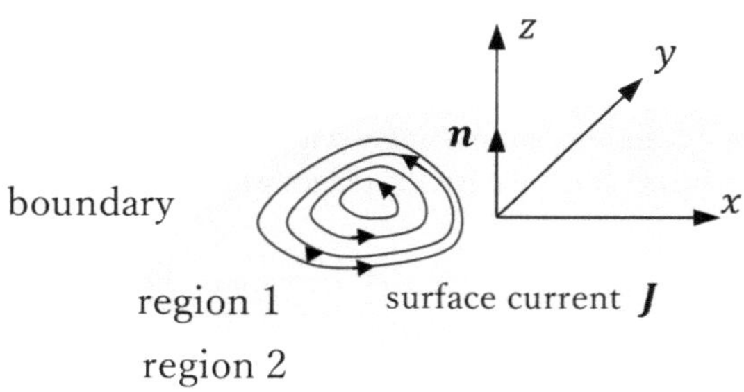

Fig. 6.25 Surface current at boundary

Besides, another boundary condition is obtained by the Coulomb gauge $\nabla \cdot \boldsymbol{A} = 0$ and $\nabla \cdot \boldsymbol{B} = 0$:

$$\begin{aligned} A_x^1 &= A_x^2 \\ A_y^1 &= A_y^2 \end{aligned} \tag{6.22}$$

That is, if a surface current is given at a plane, a difference occurs in the differential value of the magnetic vector potential normal to the plane, which is expressed by Eq. (6.21), while the value itself of the magnetic vector potential becomes continuous, as expressed by Eq. (6.22).

Now, let us return to the problem created by the coil currents $\boldsymbol{J}_s$ expressed by Eqs. (6.17) and (6.18). To find the solution of Eq. (6.20) by the method of separation of variables, by describing $\xi = (x, y)$ the solution is expressed as

$$A_{s\xi}(x, y, z) = X_{s\xi}(x)Y_{s\xi}(y)Z_{s\xi}(z) \tag{6.23}$$

Substituting this into Eq. (6.20), we have

$$\frac{X''_{s\xi}(x)}{X_{s\xi}} + \frac{Y''_{s\xi}(y)}{Y_{s\xi}} + \frac{Z''_{s\xi}(z)}{Z_{s\xi}} = 0 \tag{6.24}$$

Since each term on the left side of this equation is a function of a different independent variable, each term must be a constant. Let α, β, and γ be real numbers, and

$$\frac{X''_{s\xi}(x)}{X_{s\xi}} = -\alpha^2, \quad \frac{Y''_{s\xi}(y)}{Y_{s\xi}} = -\beta^2, \quad \frac{Z''_{s\xi}(z)}{Z_{s\xi}} = \gamma^2$$

where $\gamma = \sqrt{\alpha^2 + \beta^2}$.

By inspection of Eqs. (6.17), (6.18), (6.21), and (6.24), we get $\alpha = k = \pi/\tau$, $\beta = \pi/w$, and the magnetic vector potential of each component has the same function in terms of x and y as each surface current. Therefore, the magnetic vector potential can be expressed as

$$\begin{aligned} \dot{A}_{sx}^{\mathrm{I,II}} &= \dot{c}_{sx} e^{j(\omega t - kx)} \sin\left(\frac{\pi}{w} y\right) e^{\mp\gamma(z-h)} \\ \dot{A}_{sy}^{\mathrm{I,II}} &= \dot{c}_{sy} e^{j(\omega t - kx)} \cos\left(\frac{\pi}{w} y\right) e^{\mp\gamma(z-h)} \end{aligned} \tag{6.25}$$

where $\dot{c}_{sx}$ and $\dot{c}_{sy}$ are the unknown constants. If we apply the boundary constants at the on-board surface currents plane, we obtain the constants

$$\dot{c}_{sx} = j\frac{\mu_0}{2\gamma} J_{smx} \quad \dot{c}_{sy} = \frac{\mu_0}{2\gamma} J_{smy}$$

Thus, the magnetic vector potentials produced by the on-board surface currents are expressed as

$$\begin{aligned}\dot{A}_{sx}^{\mathrm{I,II}} &= j\frac{\mu_0}{2\gamma}J_{smx}e^{j(\omega t-kx)}\sin\left(\frac{\pi}{w}y\right)e^{\mp\gamma(z-h)}\\ \dot{A}_{sy}^{\mathrm{I,II}} &= \frac{\mu_0}{2\gamma}J_{smy}e^{j(\omega t-kx)}\cos\left(\frac{\pi}{w}y\right)e^{\mp\gamma(z-h)}\end{aligned} \tag{6.26}$$

where

$$\gamma = \sqrt{k^2+\pi^2/w^2} = \pi\sqrt{1/\tau^2+1/w^2}$$

which shows that the value of γ decreases as the pole pitch τ or the width of the conducting sheet w increases.

6.3.4 Electric Field Produced by On-Board Coil Currents and Conducting Sheet Eddy Currents

The eddy current $\dot{\boldsymbol{J}}_e$ flowing in the conducting sheet is formulated by assuming the same function as the on-board surface currents using the unknown constants $\dot{c}_{emx}$ and $\dot{c}_{emy}$ as follows:

$$\begin{aligned}\dot{\boldsymbol{J}} &= (\dot{J}_{ex}, \dot{J}_{ey}, 0)\\ \dot{J}_{ex} &= \dot{c}_{emx}e^{j(\omega t-kx)}\sin\left(\frac{\pi}{w}y\right)\\ \dot{J}_{ey} &= \dot{c}_{emy}e^{j(\omega t-kx)}\cos\left(\frac{\pi}{w}y\right)\end{aligned} \tag{6.27}$$

Thus we obtain the magnetic vector potential of the region II:

$$\begin{aligned}\dot{A}_{ex}^{\mathrm{II}} &= j\frac{\mu_0}{2\gamma}\dot{c}_{emx}e^{j(\omega t-kx)}\sin\left(\frac{\pi}{w}y\right)e^{-\gamma z}\\ \dot{A}_{ey}^{\mathrm{II}} &= \frac{\mu_0}{2\gamma}\dot{c}_{emy}e^{j(\omega t-kx)}\cos\left(\frac{\pi}{w}y\right)e^{-\gamma z}\end{aligned} \tag{6.28}$$

Both the magnetic vector potentials of Eqs. (6.26) and (6.28) create an electric field, which generates eddy currents in the conducting sheet located at $z = 0$. The electric field in the conducting sheet is produced by both the on-board coil currents and the eddy currents of the conducting sheet, which are expressed as $\dot{\boldsymbol{E}}_s = (\dot{E}_{sx}, \dot{E}_{sy})$ and $\dot{\boldsymbol{E}}_e = (\dot{E}_{ex}, \dot{E}_{ey})$, respectively. By applying the formula $\boldsymbol{E} = -\partial\boldsymbol{A}/\partial t$, we obtain the electric fields

$$\begin{aligned}\dot{E}_{sx} &= \frac{\mu_0\omega}{2\gamma} J_{smx} e^{j(\omega t - kx)} \sin\left(\frac{\pi}{w}y\right)e^{-\gamma h} \\ \dot{E}_{sy} &= -j\frac{\mu_0\omega}{2\gamma} J_{smy} e^{j(\omega t - kx)} \cos\left(\frac{\pi}{w}y\right)e^{-\gamma h}\end{aligned} \tag{6.29}$$

$$\begin{aligned}\dot{E}_{ex} &= \frac{\mu_0\omega}{2\gamma} \dot{c}_{emx} e^{j(\omega t - kx)} \sin\left(\frac{\pi}{w}y\right) \\ \dot{E}_{ey} &= -j\frac{\mu_0\omega}{2\gamma} \dot{c}_{emy} e^{j(\omega t - kx)} \cos\left(\frac{\pi}{w}y\right)\end{aligned} \tag{6.30}$$

6.3.5 Formulation of Eddy Currents in Conducting Sheet

Assuming that the thickness of the conducting sheet is small enough to neglect the skin effect, the eddy currents in the conducting sheet is given by

$$\dot{\boldsymbol{J}}_e(x, y) = \sigma d\{\dot{\boldsymbol{E}}_s(x, y) + \dot{\boldsymbol{E}}_e(x, y)\} \tag{6.31}$$

Then, we obtain the unknown coefficients $\dot{c}_{emx}$ and $\dot{c}_{emy}$ as follows:

$$\dot{c}_{em\zeta} = -\frac{j\eta_J v}{1 + j\eta_J v} e^{-\gamma h} J_{sm\zeta} \quad (\zeta = x, y) \tag{6.32}$$

where

$$\eta_J = \frac{\mu_0 \sigma d k}{2\gamma} \tag{6.33}$$

Therefore we obtain the eddy currents in the conducting sheet as

$$\begin{aligned}\dot{j}_{ex} &= \frac{\eta_J v}{1 + j\eta_J v} J_{smx} e^{j(\omega t - kx)} \sin\left(\frac{\pi}{w}y\right)e^{-\gamma h} \\ \dot{j}_{ey} &= -j\frac{\eta_J v}{1 + j\eta_J v} J_{smy} e^{j(\omega t - kx)} \cos\left(\frac{\pi}{w}y\right)e^{-\gamma h}\end{aligned} \tag{6.34}$$

The equations show that when the coil current velocity is $v < 1/\eta_J$, the strength of the eddy current in the conducting sheet tends to increase roughly proportional to the speed v, and when $v > 1/\eta_J$, the strength of the eddy current asymptotically approaches a constant value.

6.3.6 Force Densities on Conducting Sheet

In this section, we derive the electromagnetic force generated in the conductor plate by using the magnetic vector potential and eddy currents. First, the magnetic flux density $\dot{\boldsymbol{B}}_s = (\dot{B}_{sx}, \dot{B}_{sy}, \dot{B}_{sz})$ created by the on-board coil currents is expressed by the following equation.

$$\begin{aligned}\dot{\boldsymbol{B}}_s &= \nabla \times \dot{\boldsymbol{A}}_s^{\mathrm{II}} = -\frac{\partial \dot{A}_{sy}^{\mathrm{II}}}{\partial z}\boldsymbol{e}_x + \frac{\partial \dot{A}_{sx}^{\mathrm{II}}}{\partial z}\boldsymbol{e}_y + \left(\frac{\partial \dot{A}_{sy}^{\mathrm{II}}}{\partial x} - \frac{\partial \dot{A}_{sx}^{\mathrm{II}}}{\partial y}\right)\boldsymbol{e}_z \\ &= \dot{B}_{sx}\boldsymbol{e}_x + \dot{B}_{sy}\boldsymbol{e}_y + \dot{B}_{sz}\boldsymbol{e}_z\end{aligned} \tag{6.35}$$

where

$$\begin{aligned}\dot{B}_{sx} &= -\frac{\mu_0}{2}J_{smy}e^{j(\omega t - kx)}\cos\left(\frac{\pi}{w}y\right)e^{-\gamma h} \\ \dot{B}_{sy} &= j\frac{\mu_0}{2}J_{smx}e^{j(\omega t - kx)}\sin\left(\frac{\pi}{w}y\right)e^{-\gamma h} \\ \dot{B}_{sz} &= -j\frac{\mu_0}{2\gamma}\left(kJ_{smy} + \frac{\pi}{w}J_{smx}\right)e^{j(\omega t - kx)}\cos\left(\frac{\pi}{w}y\right)e^{-\gamma h}\end{aligned}$$

Then, the density of the electromagnetic force on the conducting sheet $f(x, y, t)$ [N/m^2] is expressed as the Lorentz force density

$$\begin{aligned}\boldsymbol{f}(x, y, t) &= \boldsymbol{J}_e \times \boldsymbol{B}_s \\ &= J_{ey}B_{sz}\boldsymbol{e}_x - J_{ex}B_{sz}\boldsymbol{e}_y + (J_{ex}B_{sy} - J_{ey}B_{sx})\boldsymbol{e}_z\end{aligned} \tag{6.36}$$

And, thus the levitation force density $f_L(x, y, z, t)$ [N/m^2] and the drag force density $f_D(x, y, z, t)$ [N/m^2] are given by the following equations.

$$\begin{aligned}f_L &= -J_{ex}B_{sy} + J_{ey}B_{sx} \\ f_D &= J_{ey}B_{sz}\end{aligned} \tag{6.37}$$

Example 6.3 (Force Densities on Conducting Sheet) For the model shown in Fig. 6.26, the magnetic pole pitch is $\tau = 2.0$ m, the width of the conductor plate $w = 1.0$ m, the levitation height $h = 0.25$ m, which represents the distance between the center of the on-board coil current and the center of the conductor plate, the

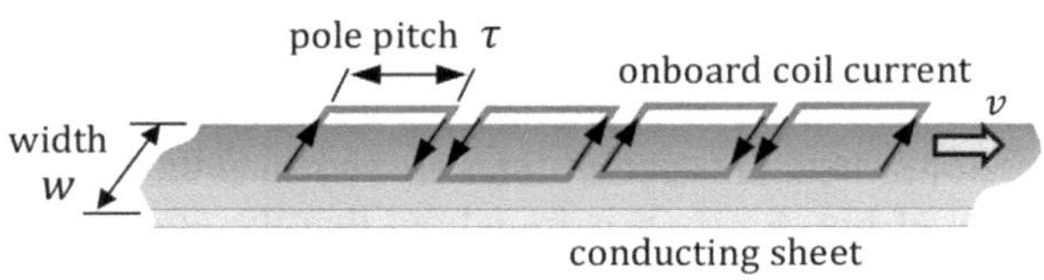

Fig. 6.26 EDS model

thickness and conductivity of the conductor plate $d = 0.01$ m, $\sigma = 3.5 \times 10^7$ S/m, respectively, and the coil magnetomotive force $NI = 200$ kA. The number of on-board cols N_M is 4. Calculate and discuss the density distributions on the conducting sheet for levitation and drag forces.

Solution The x and y components of the maximum values of coil surface currents are $J_{smx} = 200\pi$ [kA/m] and $J_{smy} = 100\pi$ [kA/m] from Eq. (6.16). The intermediate velocity v_i at which the effect of velocity begins to appear in the denominator of Eq. (6.32) is $v_i = 1/\eta_J = 2\gamma/(\mu_0 \sigma d k) \cong 10.16$ [m/s]$\cong$ 36.6 [km/h]. And, the skin depth δ [m] is given by the equation

$$\delta = \frac{1}{\sqrt{\pi \mu_0 \sigma f}}$$

If we now assume the maximum velocity is 500 km/h, then the eddy current frequency f is

$$f = \frac{\omega}{2\pi} = \frac{kv}{2\pi} = \frac{(\pi/\tau)v}{2\pi} = 34.7\,\text{Hz}$$

thus, we obtain the skin depth

$$\delta = 0.01444\,[\text{m}]$$

Since the skin depth δ is about 1.5 times the thickness of the conducting sheet ($d = 0.01$ m), the skin effect can be almost negligible. Incidentally, the magnetic Reynolds number R_m is generally defined as an index for representing the strength of the induced magnetic field. For the EDS system of this example, we have

$$R_m = \mu_0 \sigma d v = 4\pi \times 10^{-7} \times 3.5 \times 10^7 \times 0.01 \times \frac{500}{3.6} \cong 61.1 \tag{6.38}$$

The norm of the magnetic Reynolds number is 1 in terms of induced magnetic field, which means that the induced magnetic field in the conducting sheet is significantly large. In Chap. 3 we mentioned that the Reynolds number of fluid is proportional to the product of the velocity and the characteristic dimension, which may be a similar indicator to the problem here.

Figure 6.27 shows the overhead view (Fig. 6.27a) of the levitation force density distribution using Eq. (6.37) for time $t = 0$ and velocity $v = 500$ km/h, and the side view (Fig. 6.27b) of the same in the direction of travel. To make it easier to understand the distributions with respect to positional relationship, an image of the on-board coil is shown in Fig. 6.27c, though it is noted that we actually used Eqs. (6.17) and (6.18). The magnitude of the levitation force density has a maximum value of approximately 42 kN/m^2, and is positive throughout the entire region, where the contour lines are shown at the bottom of Fig. 6.27a. From the side view of Fig. 6.27b, it is found out that the effective levitation force is nearly zero

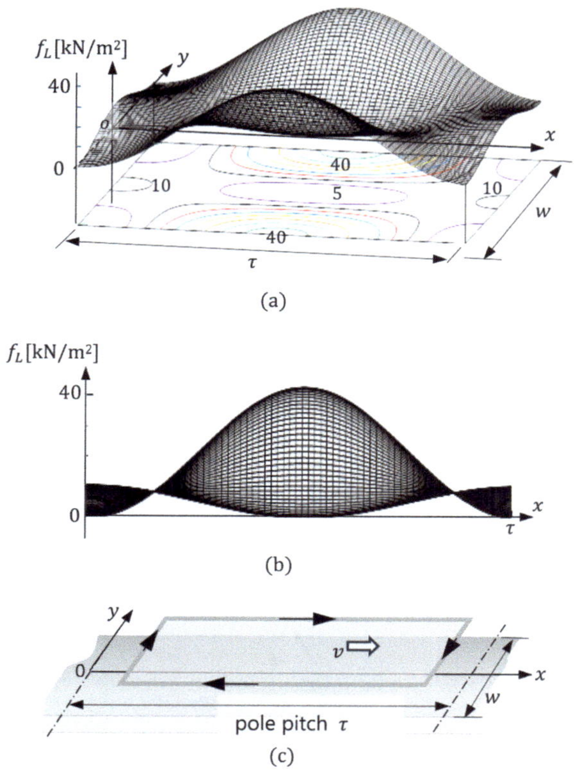

Fig. 6.27 Levitation force density on conducting sheet. (**a**) Levitation force density distribution ($t = 0,\ v = 500\,\text{km/h}$). (**b**) Side view of levitation force density ($t = 0,\ v = 500\,\text{km/h}$). (**c**) On-board coil and conducting sheet ($t = 0$)

directly below the center of the coil. That is, the levitation force is generated at both lateral ends in the conducting sheet.

Figure 6.28 shows an overhead view of the drag density on the conducting sheet using Eq. (6.37) at time $t = 0$ and velocity $v = 500\,\text{km/h}$, and a side view in the direction of travel. First, the drag has a maximum negative value ($-10\,\text{kN/m}^2$) near the front half of the on-board coils, and then takes a large positive value ($12\,\text{kN/m}^2$) in the rear half. Furthermore, as shown in Fig. 6.28b, the positive drag is greater in absolute value than the negative drag, which makes positive drag as a result. In the lateral direction, the value is large near the center, completely different from the levitation force, but in the direction of travel, it can be seen that positive and negative peaks occur near the center of the on-board coils. In other words, the drag is generated near the center just below the on-board coils.

Figure 6.29 shows the density distributions of levitation and drag at the intermediate velocity $v = 1/\eta_J = 37\,\text{km/h}$, where the effect of velocity in the denominator of Eq. (6.32) appears. As will be discussed later in the discussion of velocity characteristics, at this speed the total drag on the conducting sheet reaches at the maximum, meaning that the drag experienced by the vehicle is at its maximum. The levitation force density distributions are shifted slightly forward compared with the

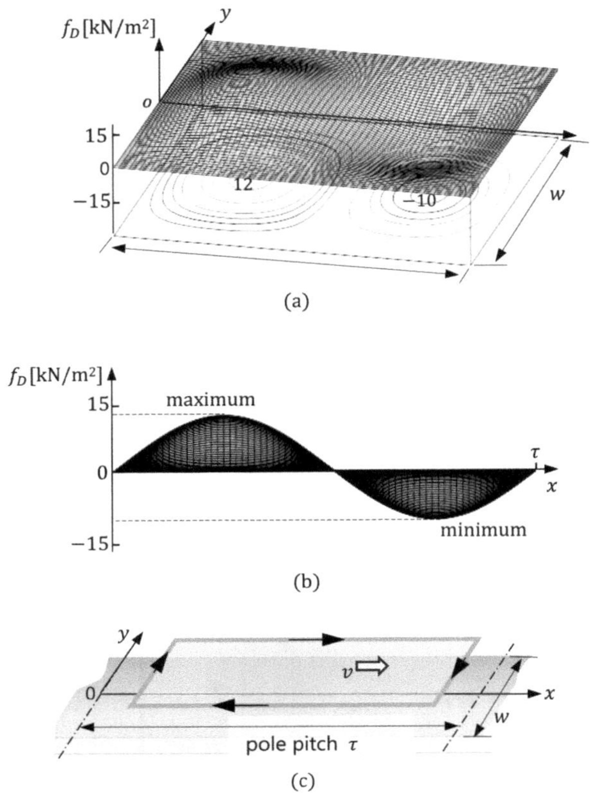

Fig. 6.28 Drag force density on conducting sheet. (**a**) Drag force density distribution ($t = 0,\ v =$ 500 km/h). (**b**) Side view of drag force density ($t = 0,\ v = 500$ km/h). (**c**) On-board coil and conducting sheet ($t = 0$)

case of 500 km/h shown in Fig. 6.27. A slight negative levitation force is generated at the rear, and if the vehicle had only the length of the magnetic pole pitch, this would result in a backward pitching moment. It can be seen that the drag force density is positive in most areas, except for a small area at the front, and the maximum value occurs at the rear.

6.3.7 Levitation and Drag Forces on Vehicle

Using a surface current density $\dot{J}$ and a magnetic flux density $\dot{B}$ expressed as complex numbers, the time average value of a force density f denoted as $\langle f \rangle$, is given by $\mathrm{Re}(\dot{J}^* \dot{B})/2$, where the asterisk "*" represents a complex conjugate. Using this relationship we obtain the levitation and drag forces averaged in time:

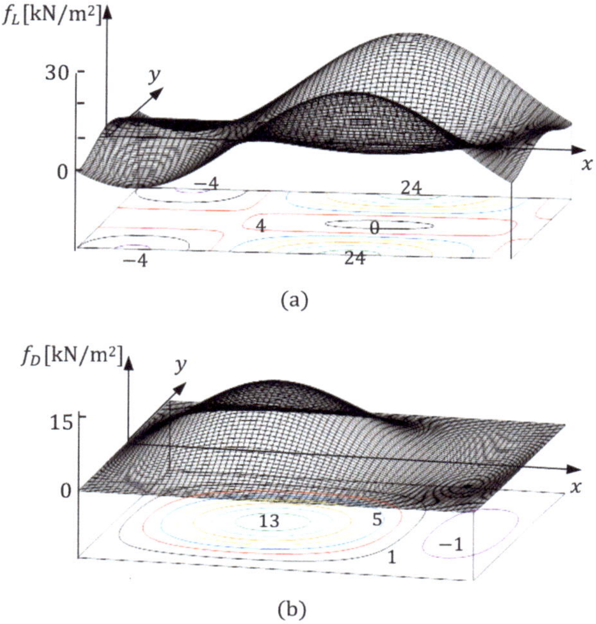

Fig. 6.29 Levitation and drag force density on conducting sheet. (**a**) Levitation force density distribution ($t = 0, \; v = 37$ km/h). (**a**) Drag force density distribution ($t = 0, \; v = 37$ km/h).

$$
\begin{aligned}
\langle f_L \rangle &= \frac{1}{2}\mathrm{Re}(\dot{J}_{ey}^{*}\dot{B}_{sx} - \dot{J}_{ex}^{*}\dot{B}_{sy}) \\
&= \frac{\mu_0}{4}e^{-2\gamma h}\frac{\eta_J^2 v^2}{1+\eta_J^2 v^2}\left\{J_{smx}^2 \sin^2\left(\frac{\pi}{w}y\right) + J_{smy}^2 \cos^2\left(\frac{\pi}{w}y\right)\right\}
\end{aligned} \tag{6.39}
$$

$$
\begin{aligned}
\langle f_D \rangle &= \frac{1}{2}\mathrm{Re}(\dot{J}_{ey}^{*}\dot{B}_{sz}) \\
&= \frac{\mu_0}{4}e^{-2\gamma h}\frac{\eta_J v}{1+\eta_J^2 v^2}J_{smy}\left(kJ_{smy} + \frac{\pi}{w}J_{smx}\right)\cos^2\left(\frac{\pi}{w}y\right)
\end{aligned} \tag{6.40}
$$

Regarding the levitation force density equation obtained here, the term related to the y-axis current J_{ey} is proportional to $\cos^2(\pi y/w)$, which shows that the levitation force density is larger at the center of the conducting sheet. On the other hand, the force density term related to the x-axis component of the current J_{ex} is distributed as $\sin^2(\pi y/w)$, which represents that the force density is larger at the lateral edges of the conducting sheet. Looking at Fig. 6.27a, we notice that there is an x-coordinate where the levitation force density has peak values at $y = 0$, and an x-coordinate where the levitation force density has peak values at $y = \pm w/2$.

Now, to find the levitation force F_L [N] and drag force F_D [N] acting on the vehicle, we assume that the vehicle has $N_M on - boardcoils$ as was shown in Fig. 6.26. The levitation and drag forces are formulated as the forces reacted from the conducting sheet with the length $N_M\tau$ [m] and width w [m].

$$\begin{aligned} F_L &= N_M \int_0^{\tau} \int_{-\frac{w}{2}}^{\frac{w}{2}} \langle f_L \rangle dxdy \\ &= \frac{\mu_0}{8} N_M \tau w e^{-2\gamma h} (J_{smx}^2 + J_{smy}^2) \frac{\eta_J^2 v^2}{1 + \eta_J^2 v^2} \end{aligned} \tag{6.41}$$

$$\begin{aligned} F_D &= N_M \int_0^{\tau} \int_{-\frac{w}{2}}^{\frac{w}{2}} \langle f_D \rangle dxdy \\ &= \frac{\mu_0}{8\gamma} N_M \tau w e^{-2\gamma h} J_{smy} \left(k J_{smy} + \frac{\pi}{w} J_{smx} \right) \frac{\eta_J v}{1 + \eta_J^2 v^2} \end{aligned} \tag{6.42}$$

Furthermore, the levitation/drag ratio is given by the following equation.

$$\frac{F_L}{F_D} = \frac{J_{smx}^2 + J_{smy}^2}{J_{smy} \left(k J_{smy} + \frac{\pi}{w} J_{smx} \right)} \gamma \eta_J v = \frac{\mu_0}{2} \sigma d v = \frac{1}{2} R_m \tag{6.43}$$

And, it was derived that the lift/drag ratio is proportional to the velocity, and also that the levitation/drag ratio is equal to half the magnetic Reynolds number R_m. Since the drag force is a magnetic force that acts as a braking force for the vehicle to travel, a large levitation/drag ratio means that less propulsion power is required to produce the necessary levitation force. ◇

Example 6.4 (Levitation Characteristics of Sheet Guideway EDS) Figure 6.30 shows the levitation characteristics based on Eqs. (6.41), (6.42), and (6.43). As the vehicle velocity increases, the levitation force asymptotically approaches a value in the high-speed region ($v \gg 1/\eta_J$), reaching approximately 106 kN at 500 km/h in this case. On the other hand, the drag force reaches a maximum value of 24 kN at the velocity of $v = 1/\eta_J \cong 37$ km/h, which appears in the low-speed region. After that velocity, it decreases inversely proportional to the velocity, and at 500 km/h it is approximately 3.5 kN. Although the levitation force and drag force change with two asymptote lines with respect to the velocity as shown in Eqs. (6.41) and (6.42), the levitation/drag ratio becomes a straight line proportional to the velocity as shown in Eq. (6.43), and at a velocity of 500 km/h

$$\frac{F_L}{F_D} = \frac{\mu_0}{2} \sigma d v = \frac{1}{2} R_m \cong 31$$

◇

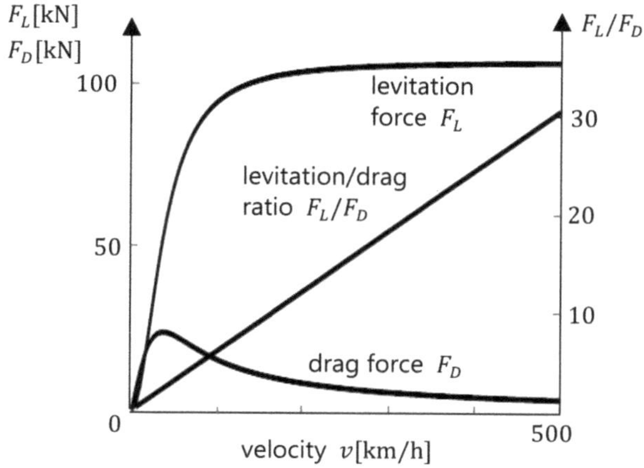

Fig. 6.30 Velocity characteristics of sheet guideway EDS

6.3.8 Eddy Currents and Electric Vector Potential

From Eq. (6.34), the equation expressed in real numbers for the eddy current in the conducting sheet is derived as

$$\begin{aligned} J_{ex} &= J_{evx}\left\{\cos(\omega t - kx) + \eta_J v \sin(\omega t - kx)\right\} \sin\left(\frac{\pi}{w} y\right) \\ J_{ey} &= J_{evy}\left\{\sin(\omega t - kx) - \eta_J v \cos(\omega t - kx)\right\} \cos\left(\frac{\pi}{w} y\right) \end{aligned} \tag{6.44}$$

where

$$J_{ev\zeta} = \frac{\eta_J v}{1 + \eta_J^2 v^2} e^{-\gamma h} J_{sm\zeta} \quad (\zeta = x, y)$$

Since a flowing current always forms a closed loop, the divergence of the eddy current of the conducting sheet satisfies the equation

$$\operatorname{div} \boldsymbol{J}_e = \frac{\partial J_{ex}}{\partial x} + \frac{\partial J_{ey}}{\partial y} = 0 \tag{6.45}$$

If the eddy current is given by using the appropriate function $T(x, y)$:

$$J_{ex} = \frac{\partial T}{\partial y}, \quad J_{ey} = -\frac{\partial T}{\partial x} \tag{6.46}$$

then, Eq. (6.45) holds true identically. The function $T(x, y)$ is called the stream function. If we now define the following equation,

$$\boldsymbol{T} = \boldsymbol{e}_z T \tag{6.47}$$

The vector function $\boldsymbol{T}$ is called the electric vector potential, which gives the relation

$$\mathrm{rot}\boldsymbol{T} = \boldsymbol{J}_e \tag{6.48}$$

Now, considering a general case of the electric vector potential $\boldsymbol{T} = (T_x, T_y, T_z)$, if a current density is written as $\boldsymbol{J}$ [A/m^2] in some region, the relationship with the magnetic field intensity $\boldsymbol{H}$ [A/m] becomes the following equation.

$$\mathrm{rot}\boldsymbol{T} = \boldsymbol{J} = \mathrm{rot}\boldsymbol{H}$$

thus, we obtain the equation

$$\boldsymbol{H} = \boldsymbol{T} - \mathrm{grad}\varphi_m$$

where φ_m is the magnetic scalar potential. This relationship is similar to the equation $\boldsymbol{E} = -\partial \boldsymbol{A}/\partial t - \mathrm{grad}\varphi$ in the case of magnetic vector potential. The methodology of analysis using the potential functions $(\boldsymbol{T}, \varphi_m)$ we derived here is a method usually called the $\boldsymbol{T} - \Omega$ approach.

Getting back to the discussion of the conducting sheet eddy currents, if we consider

$$T(x, y) = \mathrm{const.}$$

then, the total differential of $T(x, y)$ is

$$\frac{\partial T}{\partial x}dx + \frac{\partial T}{\partial y}dy = 0$$

Substituting Eq. (6.46) into this, we get the equation on the line element (dx, dy)

$$\frac{dx}{J_{ex}} = \frac{dy}{J_{ey}} \tag{6.49}$$

i.e.,

$$\frac{dy}{dx} = \frac{J_{ey}}{J_{ex}}$$

The gradient dy/dx of the curve $y(x)$ is equal to the ratio J_{ey}/J_{ex} of the x and y components of the current density, and therefore Eq. (6.49) is a differential equation representing the current streamlines.

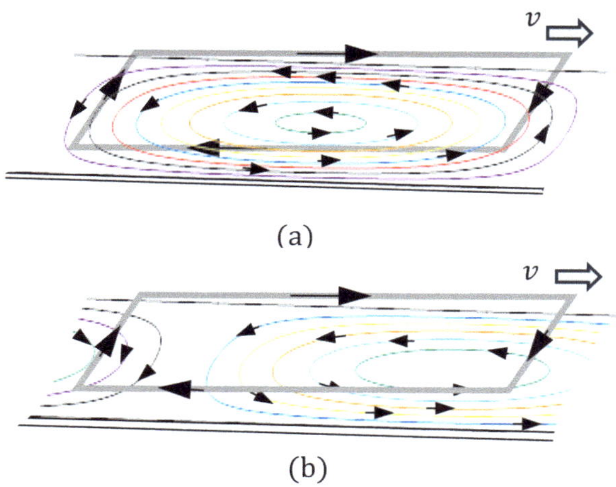

Fig. 6.31 Eddy current of conducting sheet ($t = 0$). (**a**) $v = 500$ km/h. (**b**) $v = 37$ km/h

Using Eq. (6.44), the stream function $T(x, y)$ (or, the z-component of the electric vector potential $\boldsymbol{T}$) is given by:

$$T = -NI\frac{\eta_J v}{1+\eta_J^2 v^2}e^{-\gamma h}\{\cos(\omega t - kx) + \eta_J v \sin(\omega t - kx)\}\cos\left(\frac{\pi}{w}y\right) \tag{6.50}$$

Example 6.5 (Eddy Currents of Conducting Sheet) By using Eq. (6.50) to draw the contour lines of $T(x, y)$, we get the eddy current distribution generated in the conducting sheet as shown in Fig. 6.31. Figure (a) shows the distribution at $v = 500$ km/h, from which it is found that the eddy currents are generated directly below the on-board coil current, thus the drag force is small, and the levitation force has reached its saturated value. Figure (b) shows the distribution when the speed is $v = 1/\eta_J \cong 37$ km/h, where the eddy currents are generated ahead of the on-board coil current and the eddy current density is also moderately large, so that the drag reaches its peak value, as can also be confirmed in Fig. 6.30. At lower velocity than this, the eddy current distribution moves further forward, but the eddy current density is low, so that the drag force becomes smaller.

6.4 EDS with Short-Circuited Coils on Guideway

In the previous section, we considered an EDS model in which on-board coil currents travel above a conducting sheet installed on the guideway, and analyzed the electromagnetic forces and the eddy currents in the conducting sheet. In this section, we consider a levitation method of using a short-circuited coils instead of a conducting sheet. There are mainly two types of levitation that make use of short-circuited coils: one installed on the ground of the guideway and one installed on the side wall of the guideway. We quantitatively examine both methods.

6.4.1 EDS with Short-Circuited Coils on Guideway Floor

The EDS vehicle using a conducting sheet was originally studied at MIT in the US and Warwick University in the UK. In Japan the research and development of EDS has been studied by JR, some universities and institutions, while JR has selected the method of short-circuited coils for levitation. The EDS model in Fig. 6.32 considered here is the same as the model of Fig. 6.23 except for the conducting sheet. From now on, the short-circuited coils on the guideway are referred to as the levitation coils, which require multiple coils for one pole pitch τ [m] to minimize the time pulsation of the levitation force. Then, the number of levitation coils per one pole pitch is given by $n_\tau = \tau/\tau_g$, where τ_g is the levitation coil pitch. The length and width of the levitation coil are l_g [m], w_g [m], respectively. The number of turns, the electrical resistance and the self-inductance of each levitation coil are denoted as n_g, R_g [Ω] and L_g [H], respectively.

Let's set the origin o of the coordinate system (x, y, z) to the central coordinate of one of the levitation coils of interest, derive the magnetic flux linkage equation of this levitation coil. And then we derive the electromotive force ("emf" for short)

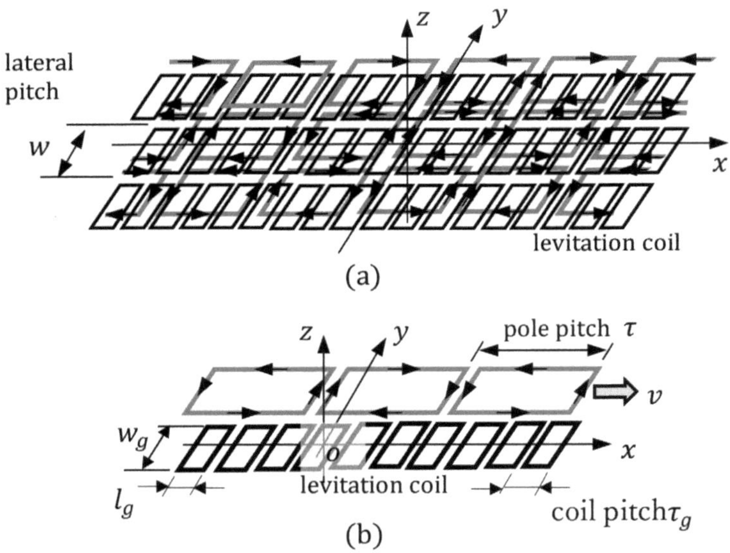

Fig. 6.32 EDS model using levitation coils on the ground. (**a**) EDS model using levitation coils. (**b**) Close-up of the model

generated in the levitation coil and the levitation coil current. If we find the time average of the electromagnetic force acting on one levitation coil, we can obtain the levitation force and drag acting on the vehicle by multiplying this by the number of levitation coils per vehicle.

Using the magnetic flux density created by the on-board coil current expressed by Eq. (6.35), the magnetic flux linkage $\dot{\psi}_{g0}$ [Wb] of the levitation coil located at the origin of the coordinate system is given by the following equation.

$$\dot{\psi}_{g0} = n_g \int_{-l_g/2}^{l_g/2} \int_{-w_g/2}^{w_g/2} \dot{B}_{sz} = -j\Psi_{g0}(h)e^{j\omega t} \tag{6.51}$$

where

$$\Psi_{g0} = \frac{2\mu_0}{\pi} n_g \frac{\tau J_{smx} + w J_{smy}}{\gamma} \sin\left(\frac{kl_g}{2}\right) \sin\left(\frac{\pi w_g}{2w}\right) e^{-\gamma h}$$

The alternating magnetic flux linkage creates the emf in the levitation coil as

$$\dot{v}_{g0} = -\frac{d\dot{\psi}_{g0}}{dt} = -\omega \Psi_{g0}(h)e^{j\omega t} \tag{6.52}$$

The voltage equation of the levitation coil is then expressed using the induced current i_{g0} as

$$R_g \dot{i}_{g0} + L_g \frac{d\dot{i}_{g0}}{dt} = \dot{v}_{g0} \tag{6.53}$$

Therefore, we obtain the induced current of the levitation coil

$$\dot{i}_{g0} = \frac{\dot{v}_{g0}}{R_g + j\omega L_g} = \frac{e^{-\theta_g}}{Z_g} \dot{v}_{g0} = -\omega \frac{e^{-\theta_g}}{Z_g} \Psi_{g0}(h)e^{j\omega t} \tag{6.54}$$

where

$$Z_g = |R_g + j\omega L_g|, \quad \theta_g = \tan^{-1} \frac{\omega L_g}{R_g}$$

In the previous section, we obtained the magnetic field and eddy currents on the conducting sheet, and the electromagnetic forces using the Lorentz force formula. Here, we employ a method based on magnetic energy to reduce the equations of forces by using the magnetic flux linkages and currents of levitation coils. Figure 6.33 illustrates a general relationship between the magnetic flux linkage of a coil ψ [Wb], the coil current i [A], and the stored magnetic energy W_m [J] in a linear system that produces electromagnetic force, such as an electromagnet. The line representing the relationship between ψ and i changes its slope with a

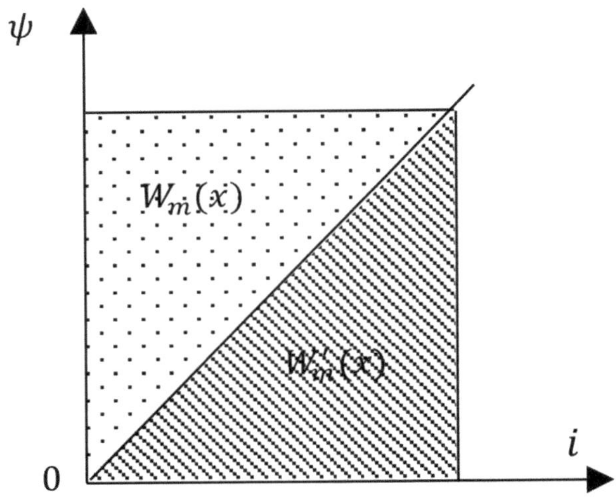

Fig. 6.33 Magnetic energy and magnetic coenergy

displacement x that tends to decrease by the resulting force. W'_m plays the role of a mathematically introduced auxiliary variable, and is called the magnetic coenergy. The rate of change of magnetic energy W_m, or magnetic coenergy W'_m with respect to the displacement x, is the electromagnetic force.

Here, we consider the magnetic energy between the on-board coil currents, where the number of coils is N_M, and one levitation coil of interest, and reduce the equation of the electromagnetic forces. The electrical terminal pair of the i-th on-board coil is composed of the current i_{vi} and the magnetic flux linkage $\psi_{vi} = M_i i_g$ that's produced from the levitation coil, where M_i is the mutual inductance between the i-th on-board coil and the levitation coil. And, the electrical terminal pair of the levitation coil of interest is the levitation coil current i_g and the magnetic flux linkage $\psi_g = \sum M_i i_{vi}$ produced from the on-board coils. Therefore, the magnetic energy of the system consisting of the on-board coil current array and one levitation coil is

$$W_m = \sum_{i=1}^{N_M} \frac{1}{2}\psi_{vi} i_{vi} + \frac{1}{2}\psi_g i_g = \psi_g i_g$$

where we used the relation

$$\sum_{i=1}^{N_M} \frac{1}{2}\psi_{vi} i_{vi} = \sum_{i=1}^{N_M} \frac{1}{2} M_i i_{vi} i_g = \frac{1}{2}\psi_g i_g$$

Thus we obtain the magnetic energy and magnetic coenergy expressed by the following equation:

$$W_m = W'_m = \psi_g i_g \tag{6.55}$$

When using the magnetic energy or the magnetic coenergy to formulate the levitation force and drag force, it is necessary to define the displacement of the system. Now, the on-board coil array coordinate system is expressed as (x_v, y_v, z_v), and the ground coordinate system of the levitation coil is (x, y, z) (Fig. 6.32). The relationship between the two is

$$(x - x_v, y - y_v, z - z_v) = (vt, 0, h)$$

The displacement of the vehicle in the x direction is expressed as

$$x_d = x - x_v = vt = \frac{\tau}{\pi}\omega t \tag{6.56}$$

The time-averaged forces per vehicle can be obtained by multiplying the time average value of the force acting between the levitation coil of interest and the on-board coil current array by the number of levitation coils per vehicle $N_M n_\tau$. Therefore, the levitation force F_L [N] and the drag force F_D [N] are given by

$$\begin{aligned} F_L &= \frac{1}{2} N_M n_\tau \langle \frac{\partial W'_m(i_g, h)}{\partial h} \rangle = \frac{1}{2} N_M n_\tau \mathrm{Re}\left\{ \frac{d\dot{\psi}^*_{g0}}{dh} \dot{i}_{g0} \right\} \\ &= \frac{\pi^2 N_M}{2\tau\tau_g} \Psi^2_{g0} \gamma \frac{L_g v^2}{Z^2_g} \end{aligned} \tag{6.57}$$

$$\begin{aligned} F_D &= -\frac{1}{2} N_M n_\tau \langle \frac{\partial W'_m(i_g, x_d)}{\partial x_d} \rangle = -\frac{1}{2} N_M n_\tau \mathrm{Re}\left\{ \frac{d\dot{\psi}^*_{g0}}{dx_d} \dot{i}_{g0} \right\} \\ &= \frac{\pi^2 N_M}{2\tau\tau_g} \Psi^2_{g0} \frac{R_g v}{Z^2_g} \end{aligned} \tag{6.58}$$

And, the levitation/drag ratio is given by the following equation.

$$\frac{F_L}{F_D} = \gamma T_g v \tag{6.59}$$

where $T_g = L_g/R_g$ is the time constant of the levitation coil. It is now found out that the levitation/drag ratio is proportional to the time constant of the levitation coil.

Example 6.6 (EDS with Short-Circuited Coils on Guideway Floor) The dimensions of the on-board coil current is the same as the model with a conducting sheet, and we use the levitation coil pitch of $\tau_g = \tau/3 = 2/3$ [m]. In the numerical example using a conducting sheet, the width of a conducting sheet was 1 m, and we determine the dimensions of the levitation coil with approximately the same amount of aluminum. As a result, the length, cross-sectional width, cross-sectional thickness, number of turns, resistance, and self-inductance of the short-circuited coil are $l_g = 0.45$ m, $w_g = 0.35$ m, $b_g = 0.08$ m, $t_g = 0.08$ m, $n_g = 50$, $R_g = 20$ mΩ,

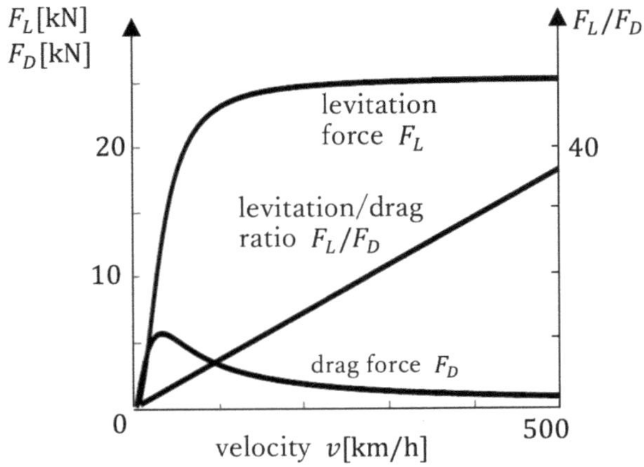

Fig. 6.34 Characteristics of EDS with levitation coils on the ground

and $L_g = 1.5$ mH, respectively. Note that the width of the coil current on the vehicle is not taken into account in the analysis, and instead only the coil installation period in the lateral direction is taken into account. In the case of a conducting sheet, the magnetic flux acts over the installation period $w = 1$ m, but if the width of the levitation coil is specified as $w_g = 0.35$ m, the magnetic flux linkage decreases by w_g/w, and the generated electromagnetic force becomes smaller by that amount. The coil width can readily be taken into account using a Fourier series, but this approach leads to finding a qualitative relationship.

Figure 6.34 shows the velocity characteristics of the electromagnetic forces. The electromagnetic force has decreased to about 1/4 compared to Fig. 6.30. This is due to the reduction in the magnetic flux linkage mentioned above. The levitation/drag ratio at 500 km/h is

$$\frac{F_L}{F_D} = \gamma T_g v \cong 3.51 \times 0.075v = 36.5$$

In an attempt to perform numerical calculations to confirm the effect of the reduction in magnetic flux linkage, the coil width was increased from w_g to the width of the conductor plate w; the results are omitted here, but the levitation and drag forces were roughly the same as in the case of a conducting sheet just as expected. Therefore, in terms of the magnitude of the generated electromagnetic forces or the levitation/drag ratio, there is no clear advantage or disadvantage between the conducting sheet and the short-circuited coil for levitation. However, in cases where eddy currents in the conducting sheet create a skin effect, the short-circuited coil has an advantage. ◇

6.4.2 EDS with Short-Circuited Coils on Guideway Sidewall

Figure 6.35 shows a model of EDS system in which the levitation coils are installed on the sidewall of the guideway. A significant feature of this system is that the short-circuited coils for levitation are shaped like a figure eight, composed of two coil elements connected in a crossed fashion, have magnetic flux linkage in the opposite direction from the on-board coil currents, resulting in a so-called null flux system, which can achieve a significant improvement in the levitation/drag ratio.

When the on-board coil is moving at the center of the two levitation coil elements, the net magnetic flux linkage to the levitation coil is 0, resulting in no electromotive force. However, as shown in Fig. 6.35b, if the on-board coil is lowered by h [m], the magnetic flux linkage to the lower coil element of the levitation coil is greater than that to the upper coil element. The currents flow in the levitation coils by the net emf's, resulting in the generation of levitation forces. The vehicle height at which the generated levitation force balances gravity is the steady-state levitation height. Figure 6.35c shows that the lateral distance between the on-board coils and the levitation coils is g [m].

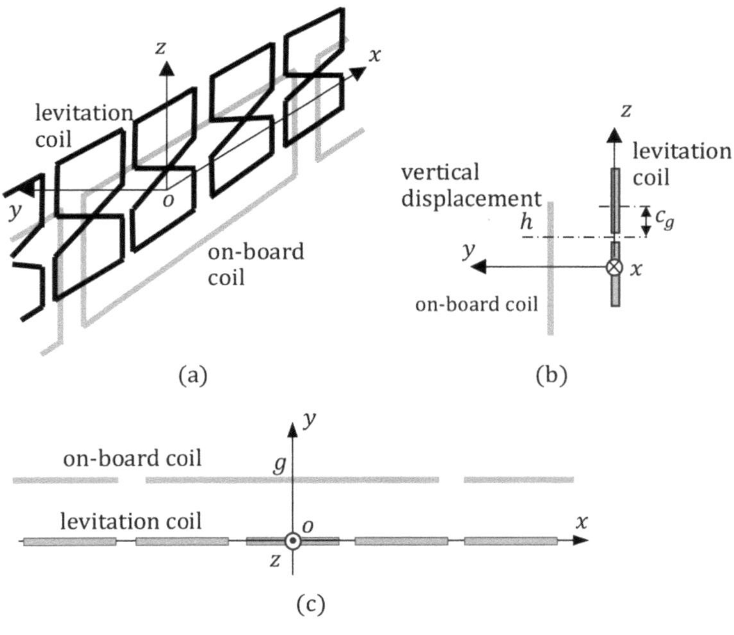

Fig. 6.35 EDS with levitation coils installed on sidewall. (**a**) Overhead view of coil arrangement. (**b**) Front view of coil arrangement. (**c**) Top view of coil arrangement

The on-board coil current array $\boldsymbol{J}_s(x, z, t) = (J_{sx}, 0, J_{sz})$ [A/m] is expressed as,

$$\begin{aligned} J_{sx} &= -J_{smx}\sin(\omega t - kx)\sin\left(\frac{\pi}{w}z\right) \\ &= \mathrm{Re}\left\{jJ_{smx}e^{j(\omega t-kx)}\sin\left(\frac{\pi}{w}z\right)\right\} \\ &= \mathrm{Re}\left\{\dot{J}_{sx}(x,z)\right\} \\ J_{sz} &= J_{smz}\cos(\omega t - kx)\cos\left(\frac{\pi}{w}z\right) \\ &= \mathrm{Re}\left\{\dot{J}_{sz}(x,z)\right\} \end{aligned} \tag{6.60}$$

where

$$J_{smx} = \frac{\pi}{w}, \quad J_{smz} = \frac{\pi}{\tau} \tag{6.61}$$

and

$$\begin{aligned} \dot{J}_{sx} &= jJ_{smx}e^{j(\omega t-kx)}\sin\left(\frac{\pi}{w}z\right) \\ \dot{J}_{sy} &= J_{smz}e^{j(\omega t-kx)}\cos\left(\frac{\pi}{w}z\right) \end{aligned} \tag{6.62}$$

To calculate the magnetic flux linkage to the levitation coil, the y-axis component of the magnetic flux density $\dot{B}_{sy}$ [T] is required, which is expressed by the following equation by substituting $y = 0$ as the y-coordinate of the levitation coils.

$$\dot{B}_{sy} = j\frac{\mu_0}{2\gamma}e^{-\gamma g}\left(kJ_{smz} + \frac{\pi}{w}J_{smx}\right)e^{j(\omega t-kx)}\cos\left(\frac{\pi}{w}z\right) \tag{6.63}$$

If the magnetic flux linkages of the upper and lower elements of the levitation coil located at the center of the x-coordinate system are denoted as $\dot{\psi}_{g0u}$ and $\dot{\psi}_{g0l}$, respectively, the total magnetic flux linkage $\dot{\psi}_{g0}$ [Wb] is given by the following equation.

$$\begin{aligned} \dot{\psi}_{g0} &= n_g\int_{-l_g/2}^{l_g/2}\left(\int_{h-c_g-w_g/2}^{h-c_g+w_g/2}\dot{B}_{sy}dz - \int_{h+c_g-w_g/2}^{h+c_g+w_g/2}\dot{B}_{sy}dz\right)dx \\ &= j\Psi_{g0}e^{j\omega t}\sin\left(\frac{\pi h}{w}\right) \end{aligned} \tag{6.64}$$

where

$$\Psi_{g0} = \frac{4\mu_0 w}{\pi}n_g\frac{kJ_{smz} + (\pi/w)J_{smx}}{k\gamma}\sin\left(\frac{kl_g}{2}\right)\sin\left(\frac{\pi w_g}{2w}\right)\sin\left(\frac{\pi c_g}{w}\right)e^{-\gamma g}$$

And, we get the emf of the levitation coil $\dot{v}_{g0}$ [V]:

$$\dot{v}_{g0} = -\frac{d\dot{\psi}_{g0}}{dt} = \omega\Psi_{g0}\sin\left(\frac{\pi h}{w}\right)e^{j\omega t} \tag{6.65}$$

The voltage equation of the levitation coil is

$$R_g\dot{i}_{g0} + L_g\frac{d\dot{i}_{g0}}{dt} = \dot{v}_{g0} \tag{6.66}$$

to obtain the levitation coil current

$$\dot{i}_{g0} = \omega\frac{e^{-j\theta_g}}{Z_g}\Psi_{g0}\sin\left(\frac{\pi h}{w}\right)e^{j\omega t} \tag{6.67}$$

where

$$Z_g = |R_g + j\omega L_g|, \quad \theta_g = \tan^{-1}\frac{\omega L_g}{R_g}$$

Noting that the levitation force in this configuration is produced in the direction that reduces the on-board coil height, i.e., the vertical displacement of vehicle h, the levitation force F_L [N], the drag force F_D, and the levitation/drag ratio F_L/F_D are given by the following equations.

$$\begin{aligned} F_L &= -\frac{1}{2}N_M n_\tau \mathrm{Re}\left\{\frac{d\dot{\psi}^*_{g0}}{dh}\dot{i}_{g0}\right\} \\ &= \frac{\pi^3 N_M}{4\tau\tau_g w}\Psi^2_{g0}\sin\left(\frac{2\pi h}{w}\right)\frac{L_g v^2}{Z_g^2} \end{aligned} \tag{6.68}$$

$$\begin{aligned} F_D &= -\frac{1}{2}N_M n_\tau \mathrm{Re}\left\{\frac{d\dot{\psi}^*_{g0}}{dx_d}\dot{i}_{g0}\right\} \\ &= \frac{\pi^2 N_M}{2\tau\tau_g}\Psi^2_{g0}\sin^2\left(\frac{\pi h}{w}\right)\frac{R_g v}{Z_g^2} \end{aligned} \tag{6.69}$$

$$\frac{F_L}{F_D} = \frac{\pi}{w}\cot\left(\frac{\pi h}{w}\right)T_g v \cong \frac{1}{h}T_g v \tag{6.70}$$

If we consider the size scale of a passenger transport vehicle the value of w is the order of π, and we approximated Eq. (6.70) by assuming

$$h < w/\pi$$

which is a valid assumption. $T_g = L_g/R_g$ is the time constant of the levitation coil. The levitation/drag ratio is proportional to the time constant T_g of the levitation coil, just as when the levitation coils are installed on the guideway floor. It is especially important that the levitation/drag ratio is inversely proportional to the vehicle displacement h, which is the distance between the center of figure-eight levitation coils and the center of the on-board coils.

The levitation/drag ratio in the case of installing levitation coils on the guideway floor was given by

$$\frac{F_L}{F_D} = \gamma T_g v$$

in which the coefficient γ is $\pi\sqrt{1/\tau^2 + 1/w^2}$, and thus it represents that the levitation/drag ratio is proportional to γ, which decreases with the system length such as the pole pitch τ or the lateral pitch w of the on-board coil. In contrast, the levitation/drag ratio is inversely proportional to the vehicle displacement h. This difference in the dependance on the levitation/drag ratio is intriguing from a view point of practical system design, and we found out now that the levitation/drag ratio of the sidewall installment of levitation coil is $1/\gamma h$ times larger than that of the floor installment. It may be noted here that these relationships were derived due to the sinusoidal formulation in either coordinate of x and y.

Example 6.7 (EDS with Short-Circuited Coils on Guideway Sidewall) The on-board coil current dimensions are the same as in the previous analysis for the floor installment of levitation coils, and the pitch of the levitation coils in the direction of travel τ_g is set to 1/3 of the magnetic pole pitch, i.e., $\tau_g = \tau/n_\tau = 2/3$ [m], as the case of the levitation coils installed on the guideway floor. The dimensions of the figure-eight levitation coils were determined so as to use approximately the same amount of aluminum as the levitation coils installed on the floor. The length, width, cross-sectional width, cross-sectional thickness, and number of turns of the upper and lower figure-eight levitation coils are set to $l_g = 0.55$ m, $w_g = 0.30$ m, $b_g = 0.08$ m, $t_g = 0.40$ m, $n_g = 40$, $R_g = 25$ mΩ, and $L_g = 2.0$ mH, respectively. Note that the mutual inductance between the upper and lower coil elements is ignored.

As mentioned above, a difference in the height value of the on-board coil displacement h has a significant effect on the levitation/drag ratio, and we consider two cases of the vehicle displacement h. Additionally, the magnetomotive force (mmf) of the on-board coil is set so as to generate a levitation force equivalent to that in the previous numerical calculation example for the floor installment. Moreover,when the vehicle displacement h is smaller, a larger on-board coil mmf is required to generate the same levitation force. The following two cases are considered.

Case 1 $h = 0.10$ m, $NI = 470$ kA
Case 2 $h = 0.05$ m, $NI = 650$ kA

Figure 6.36 shows the levitation characteristics of Case 1. The levitation/drag ratio of the floor installment was $F_L/F_D \cong 36.5$, while in this case the levitation/drag ratio is $F_L/F_D \cong 104.0$, showing a good characteristics with a levitation/drag ratio that is about three times higher. In contrast, Fig. 6.37 shows the characteristics of Case 2, where the levitation force is the same as in Case 1. However, the levitation/drag ratio reaches up to about 208 at 500 km/h, which is doubled by reducing the vehicle displacement h.

By substituting $\tau = 2$ m and $w = 1$ m of the numerical example into the equation of γ, we obtain $\gamma \cong 3.51$. Therefore, the levitation/drag ratio multiplication factor for the side wall arrangement relative to the floor arrangement is $1/\gamma h \cong 2.85$ $(= 104.0/36.5)$ when $h = 0.10$, and $1/\gamma h \cong 5.70$ $(= 208/36.5)$ when $h = 0.05$, which proves the consistency in the calculations. ◇

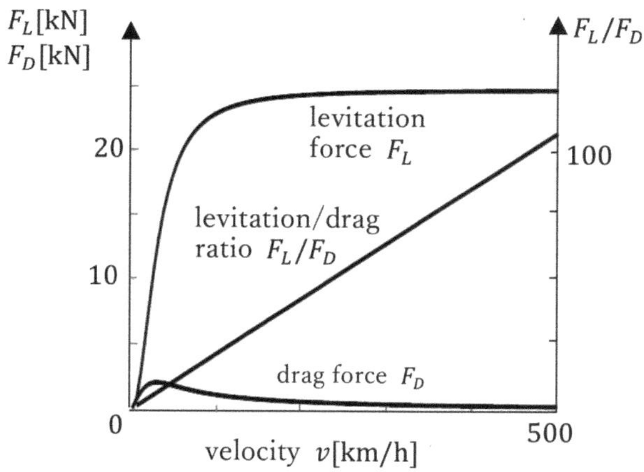

Fig. 6.36 EDS with levitation coils on sidewall ($h = 0.10$ m)

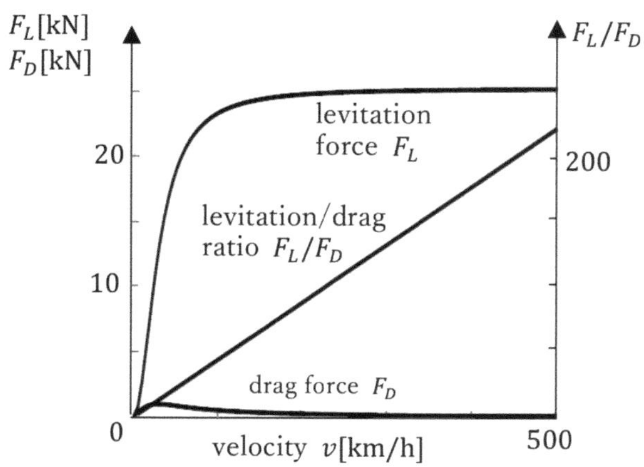

Fig. 6.37 EDS with levitation coils on sidewall ($h = 0.05$ m)

6.4.3 EDS Structures with Various Arrangement of Short-Circuited Coils

In Japan's experimental trains operated by JR, short-circuited coils have been used for the purpose of levitation rather than conducting sheet from the beginning. The on-board coil current required for the EDS, which allows for sufficient clearance, 15 to 20 cm between the vehicle and the guideway, is very large, while in the EMS the clearance is limited to 1 cm. Special metals in a superconducting state have the advantage that their electrical resistance becomes zero, and once a current flows, it continues to flow without the need for a power source to be connected to the coil. However, maintaining the superconducting state under various disturbances requires the accumulation of advanced technology over many years. It is important to note that despite the extremely large currents passing through them, no Joule loss occurs because the electrical resistance is zero, and the associated power source is used only for cooling the superconducting coils to keep them superconducting.

Figure 6.38 shows the structures of experimental vehicles developed by JR, which started with Figure (a), went through Figure (b) and has now progressed to the current Figure (c) structure. The early vehicle in Figure (a) had separate superconducting coils for levitation and propulsion, and the superconducting coils ("SC coils" for short) were located in the center of the vehicle, which was not the best structure for passenger transport in terms of space utilization. The aforementioned type with short-circuited coils laid on the guideway floor corresponds to this structure. The structure of Figure (b) is the first vehicle that utilize SC coils for both

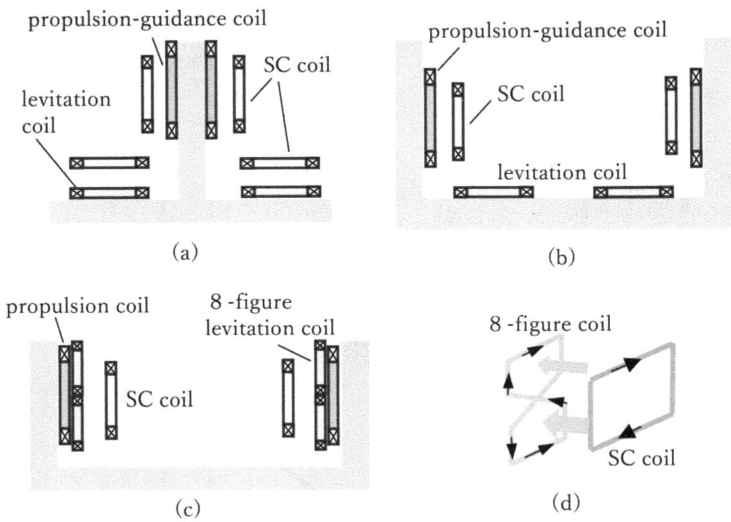

Fig. 6.38 Coil arrangement for various types of EDS. (**a**) Inverted-T-type guideway levitation coil (**b**) U-type guideway. (**c**) Sidewall coil levitation. (**d**) Principle of 8-figure coil

propulsion and levitation, which secured the space for cabin. As already described, Figure (c) is a structure introduced to improve the levitation/drag ratio, and the levitation coils and propulsion coils on the guideway are grouped together on the side wall. It should be noted that although the guidance force of the train has been omitted in the above discussion, the wiring for guidance force is connected through the guideway floor. Figure (d) shows the relationship between the SC coils and the levitation coils in Figure (c), and shows how the magnetic flux from the SC coils interlinks in opposite directions with the upper and lower coil elements.

6.5 Magnetic Field of Halbach Array PMs

Permanent magnets ("PM" for short) can be the candidate as the on-board magnet instead of SC coils for a small-scale EDS vehicle. Figure 6.39 shows an array of PMs called the Halbach array that generates much stronger magnetic field compared with an ordinary array of alternate polarity magnets. The arrows indicate magnetization vectors, and in Figure (a) the angle between adjacent magnetizations is shifted by 45°, which we call here 45° transition, while in Figure (b) it is simplified to a 90° transition, which is shifted by 90° between adjacent magnetizations. With these arrangements, the magnetic field intensity is strengthened on the lower side of the magnet array in the figure, and conversely, it can be made very weak on the opposite side.

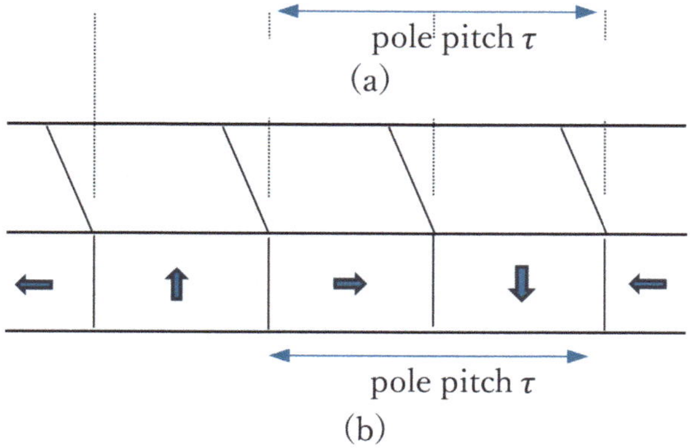

Fig. 6.39 Magnetization of Halbach array. (**a**) 45° transition. (**b**) 90° transition

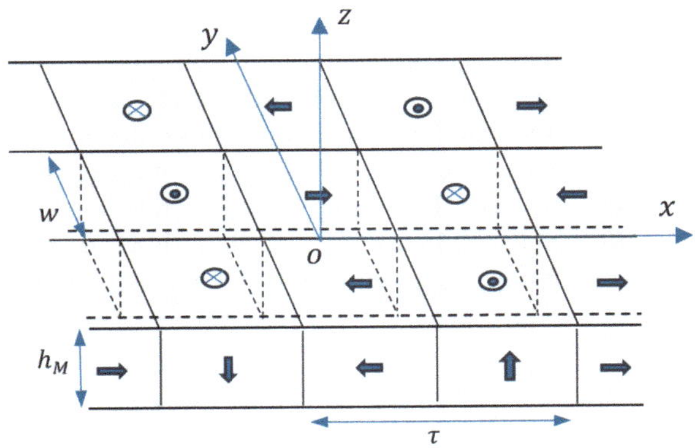

Fig. 6.40 Magnetization model of Halbach array PMs

6.5.1 Governing Equation of Magnetic Field

Figure 6.40 shows a magnetization model of the Halbach array permanent magnet. The origin of the coordinate system is set at the geometric center of the magnet piece shown in the center of the figure, and the magnet array with thickness h_M [m] is assumed to have a 90° transition of magnetic pole in the x-axis direction, with a periodic virtual arrangement set in the y-axis direction. The magnetization $\boldsymbol{M}$ [A/m] created by the magnet array has only x and z components as shown in the figure, defined as

$$\begin{aligned} \boldsymbol{M} &= (M_x, 0, M_z) \\ M_x &= M_0 \cos kx \cos \frac{\pi}{w} y \\ M_z &= -M_0 \sin kx \cos \frac{\pi}{w} y \end{aligned} \tag{6.71}$$

where the magnetizations M_x and M_z are assumed to vary in strength in the x and z directions as a sinusoidal function, and in the y direction as a virtual arrangement to have a sinusoidal function as well.

To formulate the magnetic field, we solve a boundary value problem by dividing the region of analysis into the PM region and the air region. As already mentioned, since the static magnetic field created by the magnet array satisfies the equation rot$\boldsymbol{H} = 0$, we employ the magnetic scalar potential φ_m given by

$$\boldsymbol{H} = -\text{grad}\varphi_m \tag{6.72}$$

Then, the magnetic flux density is expressed as

$$\boldsymbol{B} = \mu_0 \boldsymbol{H} + \mu_0 \boldsymbol{M} = -\mu_0 \text{grad}\varphi_m + \mu_0 \boldsymbol{M} \tag{6.73}$$

And, we have

$$\operatorname{div}\boldsymbol{B} = -\mu_0 \nabla^2 \varphi_m + \mu_0 \operatorname{div}\boldsymbol{M} = 0 \tag{6.74}$$

The change in magnetization in the PM region induces a magnetic charge density ρ_m, which is expressed by

$$\rho_m = -\operatorname{div}\boldsymbol{M} = -\frac{\partial M_x}{\partial x} = M_0 k \sin kx \cos \frac{\pi}{w} y \tag{6.75}$$

Note that the discontinuity in the z-direction of the magnetization M_z does not appear in this region, since it corresponds to the boundary surface of the analysis. As the differential operation on the boundary is not included in a region, such magnetic charge is expressed as a boundary condition discussed in the next section. Substituting the expression for the magnetic charge density into Eq. (6.74), we obtain the governing equation for the magnetic scalar potential in the region of the PMs:

$$\nabla^2 \varphi_m = -M_0 k \sin kx \cos \frac{\pi}{w} y \tag{6.76}$$

and in the region of air the governing equation is

$$\nabla^2 \varphi_m = 0 \tag{6.77}$$

6.5.2 Boundary Condition

In order to solve the magnetic field analysis as a boundary value problem, the region is defined by the xz plane shown in Fig. 6.41a. The thickness of the permanent magnet is h_M [m]. First, let the infinitesimal volume spanning regions 1 and 2 shown in Figure (b) be ΔV, the infinitesimal rectangular areas at top and bottom ΔS, and the infinitesimal thickness Δd. The boundary conditions are found by giving $\operatorname{div}\boldsymbol{B} = 0$ to the infinitesimal regions. Taking the volume integral for the volume ΔV, setting the limit for the thickness as $\Delta d \to 0$, and letting $\boldsymbol{n}$ be the normal unit vector to the infinitesimal volume surface, we obtain

$$\begin{aligned} \int \operatorname{div}\boldsymbol{B} dv &= \int \boldsymbol{B} \cdot \boldsymbol{n} dS = \int (-\mu_0 \operatorname{grad}\varphi_m + \mu_0 \boldsymbol{M}) \cdot \boldsymbol{n} dS \\ &\cong \mu_0 \left(-\frac{\partial \varphi_m^1}{\partial n} + \frac{\partial \varphi_m^2}{\partial n} \right) + \mu_0 (\sigma_m^1 - \sigma_m^2) \end{aligned} \tag{6.78}$$

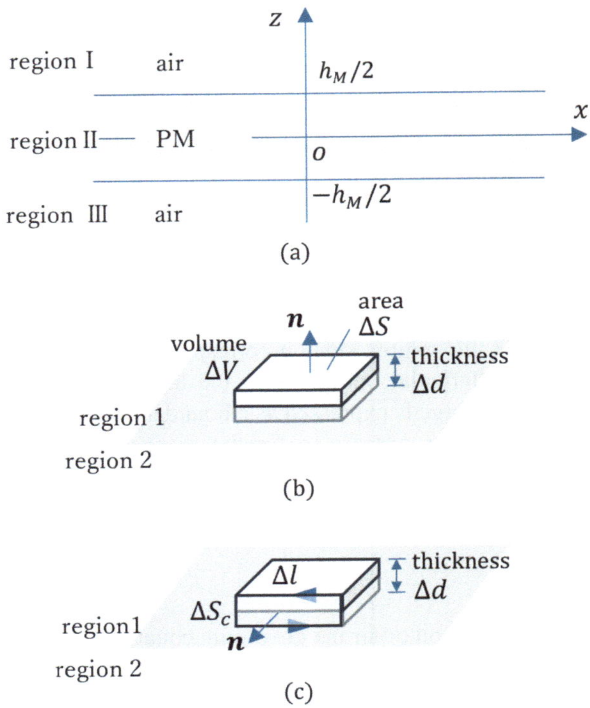

Fig. 6.41 Regions and boundary condition. (**a**) Regions for magnetic field analysis. (**b**) Neighborhood at boundary. (**c**) Neighborhood of boundary

where $\sigma_m^{1,2} = \pm \boldsymbol{M} \cdot \boldsymbol{n}$. Thus, we obtain the boundary conditions:

$$-\frac{\partial \varphi_m^{\mathrm{I}}}{\partial z} + \frac{\partial \varphi_m^{\mathrm{II}}}{\partial z} = -M_0 \sin kx \cos \frac{\pi}{w} y \quad (z = h_M/2)$$
$$-\frac{\partial \varphi_m^{\mathrm{II}}}{\partial z} + \frac{\partial \varphi_m^{\mathrm{III}}}{\partial z} = -M_0 \sin kx \cos \frac{\pi}{w} y \quad (z = -h_M/2) \tag{6.79}$$

Next, we obtain the boundary condition by setting rot$\boldsymbol{H} = 0$. If we take the surface integral of rot$\boldsymbol{H}$ over the infinitesimal side area ΔS_c across regions 1 and 2 in Figure (c), we get

$$\int \mathrm{rot}\boldsymbol{H} \cdot \boldsymbol{n} dS = \oint \boldsymbol{H} \cdot d\boldsymbol{s} \cong H_t^1 \Delta l - H_t^2 \Delta l = 0 \tag{6.80}$$

where, Δl is the tangential length in infinitesimal side area, the subscript t indicates the tangential coordinate, and we took the limit of $\Delta d \rightarrow 0$. The tangential component of the magnetic field intensity is given by $H_t^i = -\partial\varphi_m^i/\partial t$, and therefore using a unit vector $\boldsymbol{e}_t$ we have the boundary condition:

$$H_t = \boldsymbol{e}_t \cdot \boldsymbol{H} = -\boldsymbol{e}_t \cdot \nabla\varphi_m = -\frac{\partial\varphi_m}{\partial t}$$

That is, the tangential component of the magnetic field strength near the boundary surface of region $i (= 1, 2)$ can be expressed as the partial differential with respect to the tangential coordinate of the magnetic scalar potential, $H_t^i = -\partial\varphi_m^i/\partial t$. Therefore, at the boundary surface,

$$-\frac{\partial\varphi_m^1}{\partial t} + \frac{\partial\varphi_m^2}{\partial t} = -\frac{\partial}{\partial t}(\varphi_m^1 - \varphi_m^2) = 0$$

which indicates that $\varphi_m^1 - \varphi_m^2$ is constant on the boundary surface, but since the boundary area is a single layer of magnetic charge, not a double layer, the equation $\varphi_m^1 - \varphi_m^2 = 0$ holds at the boundary. Therefore, we obtain the following boundary condition:

$$\begin{aligned} &\varphi_m^{\mathrm{I}} - \varphi_m^{\mathrm{II}} = 0 \quad (z = h_M/2) \\ &\varphi_m^{\mathrm{II}} - \varphi_m^{\mathrm{III}} = 0 \quad (z = -h_M/2) \end{aligned} \tag{6.81}$$

6.5.3 Solving Governing Equations

From Eqs. (6.76) and (6.77), the governing equations are

$$\begin{aligned} &\nabla^2\varphi_m^{\mathrm{I}} = 0 \\ &\nabla^2\varphi_m^{\mathrm{II}} = -M_0 k \sin kx \cos\frac{\pi}{w}y \\ &\nabla^2\varphi_m^{\mathrm{III}} = 0 \end{aligned} \tag{6.82}$$

And, we solve the governing equations given by Eq. (6.82) with the boundary conditions given by Eqs. (6.79) and (6.81).

First, we will find the solution to the homogeneous equations expressed by equations in Eq. (6.82) by using the method of variable separation to solve the equations. We assume that the solution $\mathrm{grad}\varphi_m^{\mathrm{I}}(x, y, z)$ is expressed in the form of

$$\varphi_m^{\mathrm{I}}(x, y, z) = X(x)Y(y)Z(z)$$

then we have

$$\nabla^2 \varphi_m^{\mathrm{I}} = X''YZ + XY''Z + XYZ'' = 0$$

Rearranging this equation,

$$\frac{X''}{X} + \frac{Y''}{Y} + \frac{Z''}{Z} = 0$$

Each term on the left side is a function of only x, only y, and only z, respectively. For this equation to be valid, each term must be a constant, so that we use p, q, and r as constants to get

$$\frac{X''}{X} = -p^2, \quad \frac{Y''}{Y} = -q^2, \quad \frac{Z''}{Z} = r^2$$

where $r = \sqrt{p^2 + q^2}$. Then, using the constants a_{ij} $(i, j = 1, 2)$ the solutions are expressed as

$$X(x) = a_{11} \sin px + a_{12} \cos px$$

$$Y(y) = a_{21} \sin qy + a_{22} \cos qy$$

$$Z(z) = a_{31} e^{rz} + a_{32} e^{-rz}$$

Considering the functional forms of x and y in Eq. (6.82) and the attenuation in the z direction from the boundary position $z = h_M/2$, we may have the solutions in the regions of air φ_m^i $(i = \mathrm{I}, \mathrm{III})$ given by the forms

$$\begin{aligned} \varphi_m^{\mathrm{I}}(x, y, z) &= c_1 \sin kx \cos \frac{\pi}{w} y \, e^{-\beta(z - h_M/2)} \\ \varphi_m^{\mathrm{III}}(x, y, z) &= c_4 \sin kx \cos \frac{\pi}{w} y \, e^{\beta(z - h_M/2)} \end{aligned} \tag{6.83}$$

where c_1 and c_4 are the constants, and $k = p, \pi/w = q, \beta = r = \sqrt{k^2 + (\pi/w)^2}$.

In addition, the general solution of the inhomogeneous equation in Eq. (6.82) φ_m^{II} is written as the sum of a general solution of the homogeneous equation $\varphi_{m,g}^{\mathrm{II}}$, and a special solution to the inhomogeneous equation $\varphi_{m,s}^{\mathrm{II}}$. The special solution is obtained by the concept of eigenfunction expansion, and if we assume the solution with an unknown constant c:

$$\varphi_{m,s}^{\mathrm{II}} = c \sin kx \cos \frac{\pi}{w} y$$

then we get $c = M_0 k/\beta^2$. The general equation is written as

$$\varphi_{m,g}^{\mathrm{II}} = \left\{ c_2 e^{\beta(z-h_M/2)} + c_3 e^{\beta(z+h_M/2)} \right\} \sin kx \cos \frac{\pi}{w} y$$

And, thus we obtain the solution in the PM region with the unknown constants:

$$\begin{aligned} \varphi_m^{\mathrm{II}}(x, y, z) &= \varphi_{m,s}^{\mathrm{II}} + \varphi_{m,g}^{\mathrm{II}} \\ &= \left\{ c_2 e^{\beta(z-\frac{h_M}{2})} + c_3 e^{\beta(z+\frac{h_M}{2})} + \frac{1}{\beta^2} M_0 k \right\} \sin kx \cos \frac{\pi}{w} y \end{aligned} \tag{6.84}$$

Now that the general solution for each region has been found, let us apply the boundary conditions to find the unknown constants. Using the boundary conditions in Eq. (6.79), we get the equations for the unknown constants

$$\begin{aligned} c_1\beta + c_2\beta - c_3\beta e^{-\beta h_M} &= -M_0 \\ -c_2\beta e^{-\beta h_M} + c_3\beta + c_4\beta &= M_0 \end{aligned} \tag{6.85}$$

And, applying the boundary condition of Eq. (6.81), we get

$$\begin{aligned} -c_1 + c_2 + c_3 e^{-\beta h_M} + \frac{1}{\beta^2} M_0 k &= 0 \\ c_2 e^{-\beta h_M} + c_3\beta + \frac{1}{\beta^2} M_0 k - c_4 &= 0 \end{aligned} \tag{6.86}$$

Solving the simultaneous equations for the unknown constants give

$$\begin{aligned} c_1 &= -\frac{\beta - k}{2\beta^2}(1 - e^{-\beta h_M}) M_0 \\ c_2 &= -\frac{\beta + k}{2\beta^2} M_0 \\ c_3 &= \frac{\beta - k}{2\beta^2} M_0 \\ c_4 &= \frac{\beta + k}{2\beta^2}(1 - e^{-\beta h_M}) M_0 \end{aligned} \tag{6.87}$$

6.5.4 Lines of Magnetic Flux

Now that we have derived the solutions for the magnetic scalar potential, we can find the magnetic flux distributions in each region. The magnetic flux density in each region is given by the following equation.

$$\begin{aligned}\boldsymbol{B}^{\mathrm{I}} &= -\mu_0 \mathrm{grad}\varphi_m^{\mathrm{I}} \\ \boldsymbol{B}^{\mathrm{II}} &= -\mu_0 \mathrm{grad}\varphi_m^{\mathrm{II}} + \mu_0 \boldsymbol{M} \\ \boldsymbol{B}^{\mathrm{III}} &= -\mu_0 \mathrm{grad}\varphi_m^{\mathrm{III}}\end{aligned} \tag{6.88}$$

It should be noted here that the term $\mu_0 \boldsymbol{M}$ in the second equation is required to describe the magnetization in the whole region of PM. If this term is ignored, $\mathrm{div}\boldsymbol{B} = 0$ does not hold.

To visualize the results of the above analysis, let us draw the magnetic flux lines. In the section of the analysis for the conducting sheet eddy currents, we introduced the stream function $T(x, y)$ (or called the electric vector potential $\boldsymbol{T} = \boldsymbol{e}_z T(x, y)$) to describe the eddy current distribution. In that case, the eddy current is expressed as $\boldsymbol{J}(x, y) = \mathrm{rot}(\boldsymbol{e}_z T)$. In the field of magnetic field density $\boldsymbol{B}$, the magnetic vector potential $\boldsymbol{A}$ has a similar relationship,

$$\begin{aligned}\boldsymbol{B} = \mathrm{rot}\boldsymbol{A} &= \left(\frac{\partial A_z}{\partial y} - \frac{\partial A_y}{\partial z}\right)\boldsymbol{e}_x + \left(\frac{\partial A_x}{\partial z} - \frac{\partial A_z}{\partial x}\right)\boldsymbol{e}_y \\ &+ \left(\frac{\partial A_y}{\partial x} - \frac{\partial A_x}{\partial y}\right)\boldsymbol{e}_z\end{aligned} \tag{6.89}$$

It will be shown below that a cross-sectional view of the magnetic flux line distribution can be represented by the contour lines of the magnetic vector potential in the xz plane at coordinate $y = 0$ in Fig. 6.40.

Due to the symmetry of the magnetic field at $y = 0$, the rate of change of the magnetic flux density and magnetic vector potential with respect to the y coordinate is 0, so

$$\mathrm{div}\boldsymbol{B} = \frac{\partial B_x}{\partial x} + \frac{\partial B_z}{\partial z} = 0$$

which shows that the magnetic flux lines form closed curves in the xz plane. And, from Eq. (6.89)

$$B_x = -\frac{\partial A_y}{\partial z}, \quad B_z = \frac{\partial A_y}{\partial x} \tag{6.90}$$

If we set

$$A_y(x, z) = \mathrm{const.}$$

then

$$dA_y = \frac{\partial A_y}{\partial x}dx + \frac{\partial A_y}{\partial z} = 0$$

Substituting B_x and B_z to this equation gives

$$\frac{dx}{B_x} = \frac{dz}{B_z} \tag{6.91}$$

which shows that the contour lines of $A_y|_{y=0}$ represent the lines of magnetic flux in the xz plane.

To derive the magnetic vector potential $A_y(x, z)$, we need to integrate Eq. (6.90). However, as mentioned previously, the magnetic vector potential has an uncertainty related to the gauge, i.e., the arbitrary differentiable function χ in $\boldsymbol{B} = \mathrm{rot}(\boldsymbol{A} + \mathrm{grad}\chi)$, where rot (grad$\chi$) $= 0$. Therefore, the result of the integration may actually change depending on which term in Eq. (6.90) is used, resulting in the invalid drawing of magnetic flux lines without the continuity of contour lines. It was shown already that we formulated the magnetic filed by eliminating the uncertainty setting the Coulomb gauge $\nabla \cdot \boldsymbol{A} = 0$ in the case of employing magnetic vector potential. But, as we employed the magnetic scalar potential in this section, after examinations we have determined $A_y(x, z)$ by using the latter in Eq. (6.90):

$$A_y^{\mathrm{I}}(x, 0, z) = \int B_z^{\mathrm{I}} dx = -\mu_0 c_1 \frac{\beta}{k} \cos kx e^{-\beta(z - h_M/2)} \tag{6.92}$$

$$\begin{aligned} A_y^{\mathrm{II}}(x, 0, z) &= \int B_z^{\mathrm{II}} dx \\ &= \mu_0 \frac{\beta}{k} \left\{ c_2 e^{\beta(z - \frac{h_M}{2})} - c_3 e^{\beta(z + \frac{h_M}{2})} + \frac{M_0}{\beta} \right\} \cos kx \end{aligned} \tag{6.93}$$

$$A_y^{\mathrm{III}}(x, 0, z) = \int B_z^{\mathrm{III}} dx = \mu_0 c_4 \frac{\beta}{k} \cos kx e^{\beta(z + h_M/2)} \tag{6.94}$$

Example 6.8 (Magnetic Flux Density Distribution of Halbach Array PMs) As a numerical example, let the thickness of the PM be $h_M = 0.03$ m, the pole pitch $\tau = 0.1$ m, and the width $w = 0.15$ m. If the residual magnetic flux density is set to $B_r = 1.26$ [T], the residual magnetization is obtained as $M_r = B_r/\mu_0 \cong 10^6$ [A/m]. Then, we use the magnetization of the PM in the model as $M_0 = M_r = 10^6$ [A/m].

The results of the numerical calculations are shown in Fig. 6.42. Figure 6.42a shows a Halbach array in the x direction, with the drawing area being $-\tau \leq x \leq \tau$ and $-h_M/2 - 0.03 \leq z \leq h_M/2 + 0.01$, and Fig. 6.42b shows the A_y contour lines, i.e., the magnetic flux line distribution. The contour interval is $\Delta A_y = 0.002$ Wb/m, and the drawing range is $-0.024 \leq A_y \leq 0.024$. Note that the contour interval is

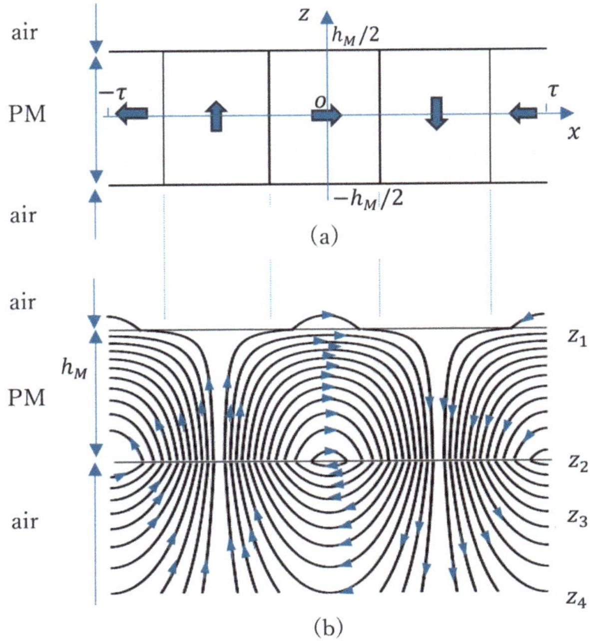

Fig. 6.42 Lines of magnetic flux of Halbach array PMs. (**a**) Magnetization of PM. (**b**) Lines of magnetic flux

twice as wide near the value of $A_y = 0$ because the line $A_y = 0$ has been deleted. The group of contour lines located in the center of the x-axis have $A_y > 0$, so that the magnetic flux lines form a right-handed vortex toward the surface of the paper. The two groups of contour lines on the left and right have $A_y < 0$ make the vortices in the opposite direction.

The z component of magnetic flux density B_z can be calculated by $B_z = (\Delta A_y)/\Delta x$ from Eq. (6.90). From the figure, it can be seen that the spacing between the densest magnetic flux lines at $z = -h_M/2$ is $\Delta x \cong 0.0025$ m, $\Delta A_y = 0.002$ Wb/m, and thus $B_z \cong 0.8$ T. A strong magnetic field is generated at $z \leq -h_M/2$, which is the front side of the PM array. In contrast, in the region $z \geq h_M/2$ on the back side of the array, the magnetic field is significantly weakened, showing the remarkable effect of the Halbach array.

In particular, at $z = -h_M/2$ the magnetic flux lines form a strong vortex center. This is expressed by the equation

$$\text{rot}\boldsymbol{B} = \text{rot}(\mu_0 \boldsymbol{H} + \mu_0 \boldsymbol{H}) = \mu_0 \text{rot}\boldsymbol{M} \tag{6.95}$$

Denoting the magnetization current as $\boldsymbol{j}_m = \text{rot}\boldsymbol{M}$ [A/m^2], we have

$$\text{rot}\boldsymbol{B} = \mu_0 \boldsymbol{j}_m \tag{6.96}$$

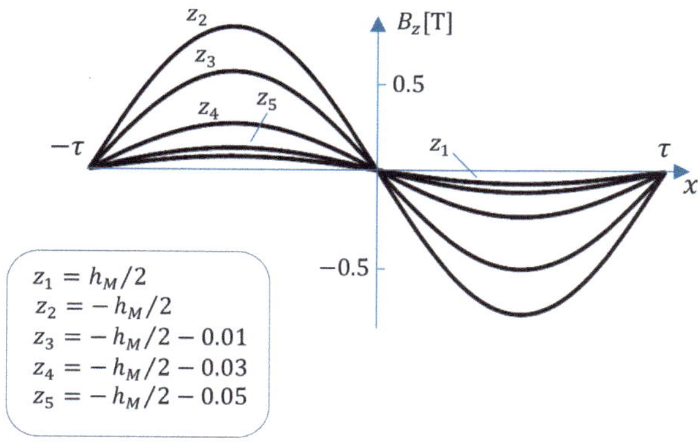

Fig. 6.43 B_z with the parameter of z coordinate (Halbach array)

It is now found out that the vortices of $\boldsymbol{B}$ are created by the magnetizing currents $\boldsymbol{j}_m$ present on the side of the PMs, and that the centers of the vortices appear in the diagram.

Figure 6.43 shows the z-component magnetic flux density B_z at $y = 0$ as a function of coordinate x, with z as a parameter. The coordinates z_1 and z_2 are the front and back surfaces of the PM array, respectively; at z_1 on the back surface, the maximum value is approximately 0.07 T at most, while at z_2 on the front surface the magnetic field is strong, reaching up to approximately 0.8 T. There appears a difference of approximately 10 times between the front and back of the PM array. As noted in the figure, the coordinates z_3, z_4, and z_5 correspond to positions away from the front surface z_2, but even at position z_3, 1 cm from the surface, the maximum value is 0.5 T, which is close to the magnetic flux density in the gap of an ordinary rotating machine. At the coordinates z_4 and z_5, the maximum magnetic flux densities are 0.25 and 0.12 T, respectively.

For comparison, the results in the case of an ordinary arrangement, i.e., an alternate polarity of PMs are shown below. The difference with the analysis of the Halbach array is that the magnetization of the x component in Eq. (6.71) is given as $M_x = 0$, and the inhomogeneous equation in Eq. (6.82) becomes a homogeneous equation. In this case, the constants of the magnetic scalar potential are obtained as follows:

$$
\begin{aligned}
c_1 &= -\frac{1}{2\beta}(1 - e^{-\beta h_M})M_0 \\
c_2 &= -\frac{1}{2\beta}M_0 \\
c_3 &= \frac{1}{2\beta}M_0 \\
c_4 &= \frac{1}{2\beta}(1 - e^{-\beta h_M})M_0
\end{aligned}
\tag{6.97}
$$

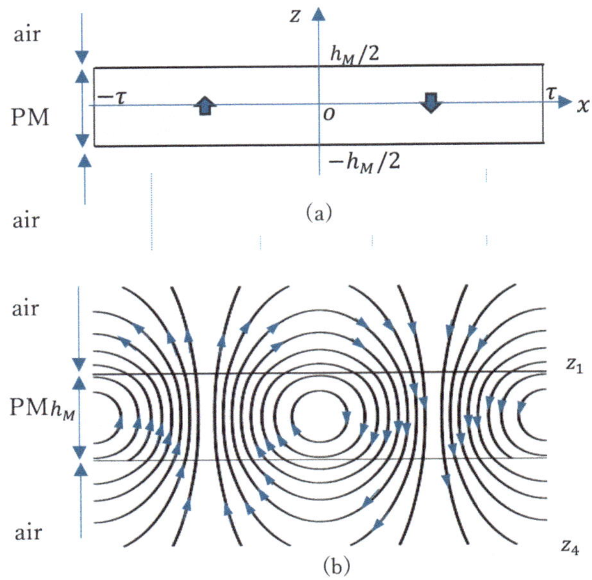

Fig. 6.44 Lines of magnetic flux of alternate array of PMs. (**a**) Magnetization of PM. (**b**) Lines of magnetic flux

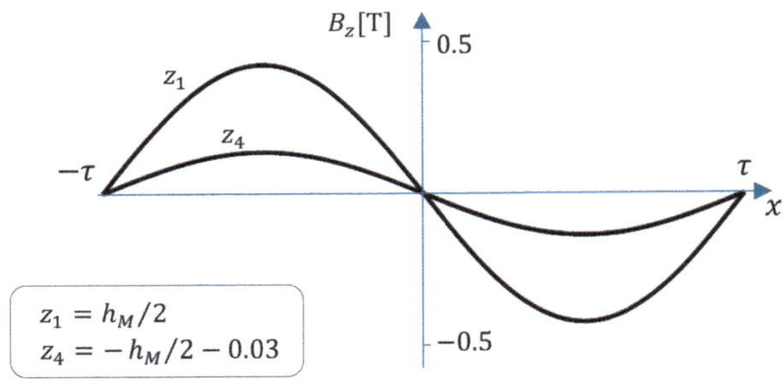

Fig. 6.45 B_z with the parameter of z coordinate (Alternate polarity array)

where we used the condition of $M_0 k/\beta^2 = 0$ in Eq. (6.84).

The magnetic flux distribution is shown in Fig. 6.44 in the range of $-\tau \leq x \leq \tau$, $-h_M/2 - 0.03 \leq z \leq h_M/2 + 0.03$. The calculated results show that the magnetic flux distribution is symmetric on both sides of the PM array, as readily expected. The contour lines are drawn with the same contour interval as in Fig. 6.42. The magnetic field on the front side of the magnet array (lower part of the figure) is half the value in the case of the Halbach array. Since the distribution is symmetrical from top to bottom, the magnetic field on the back side of the array (upper part of the figure) has the same value as the front side, even though the back side field is not required for normal use. Furthermore, the vortices of magnetic flux lines occur in the center of the PM array, which may represent the efficiency of creating magnetic field where expected. Figure 6.45 shows the distribution of magnetic flux density B_z for the coordinates z_1 and z_4 for comparison with Fig. 6.43. At z_1 on the back surface of the magnet array, the alternate polarity array has a maximum magnetic flux density of about 0.42 T, while the Halbach array suppresses the magnetic field to a very weak value of about 0.07 T. For example, at the coordinate z_4, the magnetic field is approximately only 0.14 T, which is half the magnitude of 0.25 T in the Halbach array. ◇

Above, we have conducted a detailed analysis of the Halbach array PMs, which is suitable for EDS of medium-sized or smaller vehicles, and have examined the phenomenon based on the analysis. This array methodology has high practical value in that it generates a magnetic field twice as large as a normal array on the front side of the array, while suppressing the magnetic field to a negligible level on the opposite side. In the Halbach array, the adjacent magnetizing currents of the PMs have the same direction on the area of the front side surface, while on the back side the adjacent PMs create magnetizing currents in the opposite directions so as to cancel each other's field. Thus, the magnetizing currents are located on the front side surface, but on the back side the magnetizing currents existence is largely canceled.

6.6 Structure and History of Maglev Vehicle

In the EMS, the magnetic flux density is maximized for the current flowing in the coil by utilizing the high magnetic permeability of ferromagnetic materials, and the required magnetic force is obtained. However, the gap length, which is the clearance between the vehicle and the guideway, cannot be made too large. As the gap length increases, the magnetic field in the gap not only diffuses in the width direction, but also the system cannot generate an electromagnetic force with high efficiency of the electric circuits. Therefore, the practical upper limit of the clearance is about 10 mm, and the vehicle velocity is determined in relation to the cost of guideway maintenance for safety. If there are irregularities such as protrusions on the surface of the guideway, it may be extremely difficult to respond to them because of the sampling time of control and the response speed. The magnetically levitated vehicles are called Maglev for short. The Maglev vehicle with EMS, however, has the advantages that it uses low-cost electromagnets, and that the vehicle can levitate even when the velocity is 0.

The Maglev vehicle with EDS using SC coils is highly efficient, making it possible for the vehicle to run with a large clearance between the guideway and the vehicle. A vehicle with sufficient clearance is advantageous in terms of guideway maintenance and safety. However, the Maglev vehicles with EDS require the use of expensive SC coils, and the supporting wheels are required at speeds below the levitating velocity.

Furthermore, in the EMS a stable magnetic spring and magnetic damping for vertical movement are secured by the control system design. On the other hand, in the EDS, the combination of a superconducting coil and a levitation coil alone does not provide dynamic stability, but positive magnetic damping is provided by

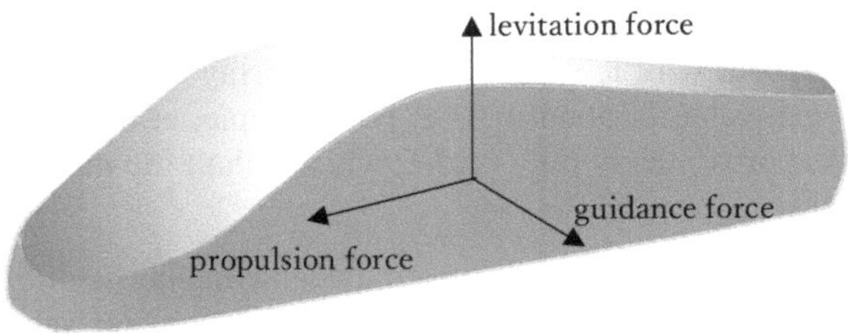

Fig. 6.46 Three fundamental forces of Maglev

the eddy currents flowing in the conductive vessel of the SC coils. When PMs are used, the damping may be available from the slight conductivity of the PMs, but it is also possible to address this problem by adding a damper coil.

For these reasons, the EDS Maglev is suitable for vehicles that travel at high speeds, while the EMS Maglev is suitable for vehicles that travel at low to mid-speed ranges.

In addition, the EDS Maglev vehicles that use short-circuited coils for levitation have the great advantage that, unlike EDS vehicles with conducting sheets, the time harmonic magnetic field is generated by the currents of the short-circuited coils as seen from the vehicle. Thus, it is possible to collect power without contact by simply placing collecting coils on the vehicle, which is called a linear generator.

In addition to the levitation force, another two forces are required for non-contact running of Maglev vehicles: propulsion force and guidance force (Fig. 6.46). To achieve this, a linear motor is used to generate the propulsion force, and another connections of coils between the both sides of the vehicle, propulsion coils or levitation coils on the guideway to generate the guidance force.

6.6.1 Propulsion Force

Linear motors are mainly classified into Linear Synchronous Motors (LSM) and Linear Induction Motors (LIM) (Fig. 6.47). LSM is a combination of an armature connected to the three-phase power source and the field using SC coils or PMs, while LIM is composed of a primary windings connected to the three-phase power source and the secondary made of conducting plate where eddy currents are produced. The secondary conducting plate has a steel plate made of silicon steel combined on the back side to induce magnetic flux. In the diagram, when three-phase voltages, labeled U-phase, V-phase, and W-phase, are applied to the coils connected in series in each phase (Fig. 6.47c), traveling magnetic field is generated. A variable-voltage variable-frequency (VVVF) inverter is generally used as the three-phase power source.

In the case of the EDS using SC coils, the SC coils are also used to generate propulsion forces to construct the LSM. The magnetic field generated by the enormous mmf of the SC coils is strong, so unlike normal rotating machines, an iron core, which enables the generation of large magnetic flux, is not used. Even if

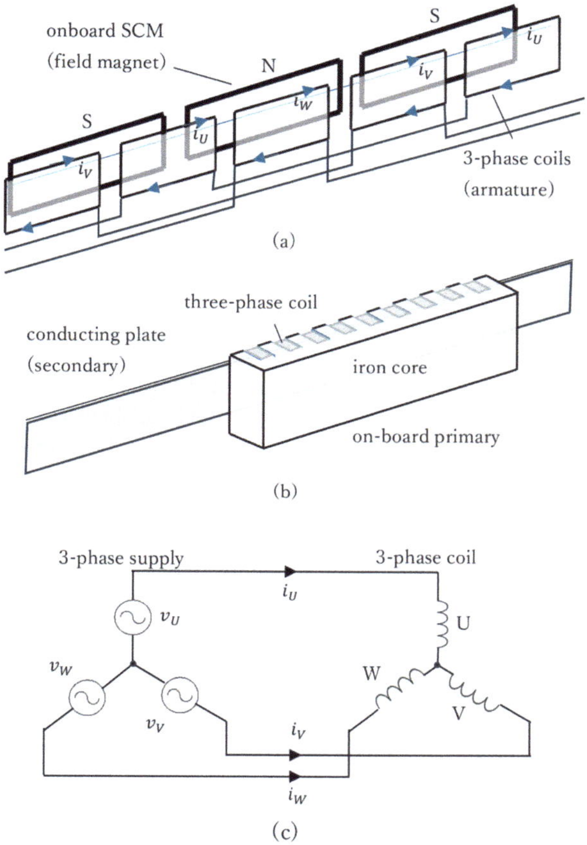

Fig. 6.47 Propulsion for Maglev. (**a**) LSM. (**b**) LIM. (**c**) LIM propulsion coil connected with 3-phase supply

an iron core was used, the extremely strong magnetic field of the SC coils may cause magnetic saturation in the iron core, and the magnetic field generated by the iron core itself becomes saturated, and in that case the magnetic permeability ultimately is the same as that of air. The air-cored LSMs using SC coils are suitable for super-high speed vehicles because, for a large clearance between the vehicle and the guideway, sufficiently large induced emf is generated in the armature coils due to the strong magnetic field of the SC coils. And the electrical energy from the power source is efficiently converted into mechanical energy for propulsion. Incidentally, the array of SC coils on the vehicle is equivalent to magnets, so they are also called superconducting magnets (SCM). In the LSM using SCMs, the propulsion force can be controlled by controlling the magnitude of the current in the propulsion coils to match the position of the magnetic poles of the SCM, marked N and S in the diagram.

In the case of the EMS, as the clearance becomes larger, leakage flux increases and the magnetic flux component that effectively creates an attractive force decreases significantly, so the clearance between the vehicle and the track is limited.

Also, unlike the EDS, there is no element that generates a strong magnetic field such as SC coils, so that LSM or LIM using an iron core is used. Furthermore, even if attempts are made to increase the clearance, LIM uses a conducting plate on the secondary, so the eddy currents generated in the conducting plate cannot be expected to generate a strong magnetic field, making effective energy conversion between the primary and the secondary difficult, and the clearance is limited to a small size, i.e, about 1 cm at most. The same is true for LSMs, and when a normal-conducting electromagnet is used for the field magnet, the clearance must be small to make the energy conversion between the armature input power and the propulsion power effective.

Figure 6.47 shows an example of the LSM that uses SCMs for the field magnet and the LIM that uses a conducting plate on the secondary. However, since the SCM is not only expensive but also needs to be kept at an extremely low temperature, it is necessary to place it on the vehicle, and therefore the armature coils of the LSM are installed on the guideway. Even in small-scale vehicles that use PMs instead of SCMs, laying PMs on the guideway requires excessive maintenance costs, so it is also necessary to load the PMs on the vehicle. As for the structure of the LSM armature windings, because the LSM armature coil must be laid over a long distance, one method to reduce the cost is a method called the single-layer mono-pole arrangement, in which rectangular coils with 120° coil pitch, single layer, mono-pole arrangement are installed in as shown in the figure. However, if the distribution of the moving magnetic field in the longitudinal direction deviates from a sine wave and contains harmonic components, the electromagnetic forces acting on the SC coils contain time fluctuations, which can lead to a state called the quench that makes the superconducting state lost. Therefore, the Yamanashi test line in Japan has 240° coil pitch, double-layer, mono-pole arrangement, which allows some overlap of the coils to improve the magnetic field distribution. The improvement of superconducting stabilization technology against the vibration of the SC coils has also been remarkable. The significant feature of the air-cored LSM is that it allows for high-speed traveling with high efficiency while having a large clearance. In addition, the electric power is supplied to the LSM armatures where the vehicle is running on the guideway by dividing the armature into appropriate sections. This allows for rational power supply to the propulsion coil, which is called the feeding system. In order to prevent the strong magnetic field of the SCM from affecting the passengers, the SCMs are installed only at the vehicle joints except for the passenger compartments, and the passenger entrance is magnetically shielded.

On the other hand, the conducting plate of the LIM has the advantage of being low cost, so the secondary is installed on the guideway and the primary is placed on the vehicle. However, the length of the primary is limited to the length of the vehicle at most, then the moving magnetic field is generated only within that finite length. In this case, a sudden change in the magnetic field distribution occurs at the entrance and exit of the primary, and the unevenness of the eddy current distribution on the conducting plate becomes noticeable as the velocity increases. As a result, thrust, power factor, and efficiency also decrease significantly as the velocity increases. This phenomenon is known as the end effect. Therefore, the appropriate running speed of the vehicle with the LIMs is said to be a maximum of 200–300 km/h. Various proposals have been made to reduce the end effect so far.

6.6.2 Guidance Force

There are two ways to generate a guidance force of Maglev without contact: employing the attractive force generated between the electromagnets on the vehicle and the ferromagnetic material on the guideway, and employing a null-flux connection that uses the difference in the magnitude of the induced emf caused by the SCMs, which is generated in the propulsion coils or levitation coils located on both sides of the vehicle. In Fig. 6.48a, the iron core on the guideway facing the electromagnets is shaped like the letter U, and when the central axis is misaligned, the restoring force is used as a guidance force. In other words, when the vehicle deviates in the guidance direction, the attractive force decreases and the clearance increases, so the coil current increases due to the control system, and the attractive force increases, resulting in a correction in the guidance direction. In the case of Fig. 6.48b, the guidance force is generated independently of the levitation force, and the guidance force is controlled by detecting the deviation in the vehicle.

The null-flux connection is used for the EDS Maglev vehicles, while in the case of propulsion coils the connections are made for high voltages, requiring high-voltage insulating materials. Figure 6.35 described the sidewall levitation vehicle on the Yamanashi test line, where the levitation coils on both sides of the guideway have null-flux connections as shown in Fig. 6.49. And, thus the levitation coils also generate guidance forces.

To provide a simple physical explanation of how guidance forces are generated, let us consider two pairs of levitation coils, arbitrarily selected on the left and right sides of the guideway. When a vehicle shifts laterally, even without null flux connections, the repulsive force from the lower coil elements, which has a larger area facing the SCMs in the figure-eight levitation coils, becomes stronger on the

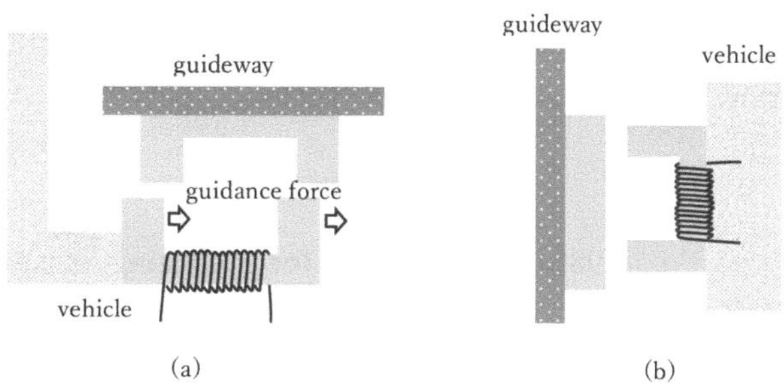

Fig. 6.48 Guidance force by electromagnets. (**a**) Levitation and guidance forces. (**b**) Guidance force

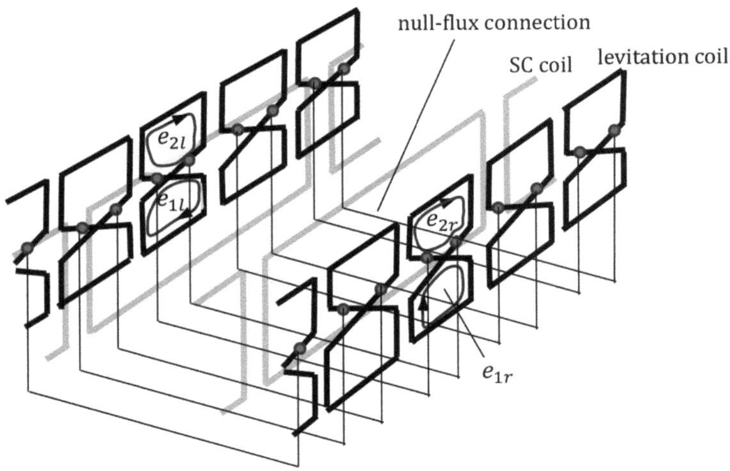

Fig. 6.49 Guidance system using null-flux connection of levitation coils

approaching side, and conversely, the opposite side becomes weaker, so a guidance force is exerted. However, the null-flux connections strengthen this effect in the following way.

The electromotive forces generated by the movement of the vehicle's SCMs in the upper and lower coil elements located on the left side of the guideway of the figure-eight levitation coil, which is divided into upper and lower elements and connected as shown in the figure, are respectively e_{2l} and e_{1l}. Similarly, the electromotive forces on the right side are e_{2r} and e_{1r}. Here, the clockwise direction on the figure is the positive direction of voltages and currents, and the direction in which the right-hand thread advances is the positive direction of the magnetic field. Note that for levitation force to be generated, as shown in Fig. 6.35b, the height center of the SCMs must be located below the center of the levitation coils, and the following relations must hold.

$$e_{1l} > e_{2l}, \quad e_{1r} > e_{2r} \tag{6.98}$$

If the vehicle runs at the lateral center of the levitation coils, then

$$e_{1l} = e_{1r}, \quad e_{2l} = e_{2r} \tag{6.99}$$

In other words, Eq. (6.98) is the condition for generating levitation forces, and because the upper and lower elements of the levitation coils are connected, currents flow in the directions according to e_{1l} and e_{1r} in the left and right levitation coils, respectively. If the vehicle is not running at the lateral center of the guideway, the emfs on the side where the vehicle has come closer increase than the other side. And, if the vehicle is displaced on the right, we have

$$e_{1l} < e_{1r}, \quad e_{2l} < e_{2r} \tag{6.100}$$

This emphasizes the effect of current flowing from the right-hand levitation coil to the left-hand levitation coil through the null-flux connection, making the repulsive force between the SCM and levitation coil greater on the right than on the left. The null-flux connection is an effective way of generating the electromagnetic stiffness in the guidance system of the vehicle. In this way, the levitation coils on both sides of the guideway function not only as levitation but also as guidance.

When using figure-eight levitation coils for the EDS Maglev, as already mentioned, if the center height of the SCMs and the center of the levitation coils are aligned, no levitation force or drag is generated. From the viewpoint of power consumption, it is not rational to run levitated at low speeds where the drag force is at its maximum. Accordingly, in low speeds, the vehicle runs on wheels using the thrust of the LSM at the height where the center of the SCMs and the center of the levitation coils are aligned. When the vehicle reaches a certain speed where sufficient guidance force is generated, drag force becomes small, the operation is switched to levitated running. This is similar to how an airplane starts running on its wheels, retracts the wheels at a speed where the necessary lift is generated, and takes off.

6.6.3 History of Maglev Vehicles

Magnetic levitation technology dates back to the German Hermann Kemper, who successfully conducted an experiment on EMS in 1933, and various active research and development efforts have taken place in Germany. In 1966, Americans James Powell and Gordon Danby proposed the concept of an EDS Maglev vehicle that use a strong magnetic field created by SC coils. The world's first commercial vehicle using magnetic levitation was the Air-Rail Link, a practical line for airport access in Birmingham, England, which opened in 1984 and operated until 1995 with a line length of 600 m and a levitation height of 15 mm using the short primary LIM propulsion.

Meanwhile, in Germany, research and development of Transrapid 01 train was started by Krauss-Maffei in 1969, and other research organizations such as MBB and Thyssen Henschel were integrated into it in 1978. In 1983 the experiments with Transrapid 06, which used EMS for levitation and LSM for propulsion with the armature windings on the guideway, were carried out on a 31.5 km long track in Emsland, and a speed of 412 km/h was recorded in 1988. The test vehicle Transrapid 09, introduced in 2007, was capable of non-contact power collection at all speeds. Although no practical line has ever been built in Germany, a 30.5 km practical line with a clearance of 10 mm was completed in Shanghai, China, in 2002 for airport access. In Japan, the R&D of the EMS Maglev vehicle HSST was started by Japan Airlines in 1972 for the purpose of airport access, and was succeeded by Chubu HSST Development Corp. in 1991. The company started operations of Linimo in 2005, and currently operates in Aichi Prefecture as a three-car train with the short primary LIM at a maximum speed of 100 km/h on a route length of 8.9 km. In South Korea, an experimental EMS vehicle was built in 1986, and a LIM-propelled vehicle was unveiled at the 1993 World Expo. Since 2016, it has been operating at a maximum speed of 110 km/h on a commercial route length of 6.1 km.

Research and development of EDS Maglev vehicles using SCMs was carried out in the United States in the 1970s, mainly at MIT, and a 1/25-scale test vehicle called Magneplane was built with a 116 m long track. In Germany, a consortium consisting of AEG-Telefunken, BBC, and Siemens built a test vehicle called EET-01 using SCMs, which recorded a speed of 401 km/h on the Erlangen circular track. However, in 1979, the West German government stopped funding the EDS vehicles after conducting a comparative study of the technical, operational, and environmental aspects of the EDS Maglev vehicles with the EMS Maglev vehicles.

In Japan, research on the EDS vehicles using SCMs began in 1962 at the Railway Technical Research Institute, and in 1972 they successfully ran the test vehicle LSM200 with the LSM propulsion and the test vehicle ML100 with the LIM propulsion. In 1977 the Miyazaki test center was opened, where experiments with the EDS vehicle ML500 were carried out, setting a speed record of 517 km/h in 1979. After several experimental vehicles were used, the site was moved to the Yamanashi Maglev test center, where in 1997 the test vehicle MLX01 with two cars succeeded in manned running at 531 km/h and unmanned running at 550 km/h. In 2015 the test vehicle L0, developed by Central Japan Railway Co. as a commercial vehicle, recorded a speed of 603 km/h. The construction schedule for future operational lines has been delayed by about seven years from the original plan due to a political reason, and the 290 km line between Tokyo and Nagoya, with a maximum speed of 505 km/h, is expected to be completed after 2034, and the 438 km line to Osaka is probably planned for after 2050.

Meanwhile, NASA in the United States is developing a rocket launcher that uses the EDS system with the Halbach array PMs. This will accelerate the spacecraft to a speed of 950 km/h by Maglev using LSM propulsion, and then ignite the engine.

Bibliography

1. D. Schieber, *Electromagnetic Induction Phenomena* (Springer, 1986)
2. I. Imai, *Fluid Dynamics and Complex Analysis* (in Japanese) (Nippon-hyoron-sha Co. Ltd, 1981)
3. Railway Technical Research Institute, *Linear Motor Car Maglev* (in Japanese) (Seibunsha Co. Ltd, 1989)
4. S. Hasebe, Y. Kano, Relation of $T - \Omega$ Method and a new current model in magnetostatic fields analysis (in Japanese). T. IEE Jpn. **110-A**(9), 555 (1990)

5. S. Fujiwara, Characteristics of EDS magnetic levitation having ground coils for levitation arranged on the side wall (in Japanese). T. IEE Japan **106-D**(5), 439 (1988)
6. H.H. Woodson, J.R. Melcher, *Electromechanical Dynamics*, Part 1 (Wiley, 1968)
7. S. Fujiwara, T. Fujimoto, Characteristics of combined levitation and guidance EDS Maglev system (in Japanese). T. IEE Jpn. **112-D**(5), 459 (1992)
8. T. Murai, Characteristics of LSM combined propulsion, levitation and guidance (in Japanese). T. IEE Japan **114-D**(7/8), 746 (1994)
9. S. Nonaka, T. Sakamoto, Three-dimensional analysis of dynamic eddy current of a cryostat outer vessel in a superconducting magnetically levitated vehicle. Electr. Eng. Jpn. **106**(5), 82 (1986)
10. J.R. Powell, G.R. Danby, High speed transport by magnetically suspended trains, in *ASME Winter Annual Meeting, 66-WA/RR-5* (1966)
11. H.H. Kolm, R.D. Thornton, Y. Iwasa, W.S. Brown, The magneplane system. Cryogenics **15**(7), 377 (1975)
12. C. McFadden, The rise and fall of Transrapid. https://interestingengineering.com/innovation/rise-and-fall-of-transrapid-maglev-train
13. T.R. Eastham, High-speed rail: another golden age? Sci. Am. **273**(3), 100 (1995)
14. A. Werske, Transrapid Magnetschwebebahn in Deutschland. https://www.hochgeschwindigkeitszuege.com/deutschland/transrapid.php
15. RTRI, History of superconducting maglev (in Japanese). https://www.rtri.or.jp/rd/maglev/history.html
16. JR Central, "SC Maglev". https://scmaglev.jr-central-global.com/
17. R.F. Post, D.D. Ryutov, The Inductrack approach to magnetic levitation, in *Proc. MAGLEV2000* (2000), p. 15
18. J.V. Masi, Overview of Halbach magnets and their applications, in *Proc. Electrical Manufacturing & Coil Winding Conference* (2010)

MIX
Papier aus verantwortungsvollen Quellen
Paper from responsible sources
FSC® C105338

If you have any concerns about our products,
you can contact us on
ProductSafety@springernature.com

In case Publisher is established outside the EU,
the EU authorized representative is:
Springer Nature Customer Service Center GmbH
Europaplatz 3, 69115 Heidelberg, Germany

Printed by Libri Plureos GmbH
in Hamburg, Germany